D0844776

Quality and Reliability of Telecommunications Infrastructure

TELECOMMUNICATIONS

A Series of Volumes Edited
by Christopher H. Sterling

Quality and Reliability of Telecommunications Infrastructure

Edited by

William Lehr
Columbia University

LEA LAWRENCE ERLBAUM ASSOCIATES, PUBLISHERS
1995   Mahwah, New Jersey   Hove, UK

Lawrence Erlbaum Associates, Inc., Publishers
10 Industrial Avenue
Mahwah, New Jersey 07430

Library of Congress Cataloging-in-Publication Data

Quality and reliability of telecommunications infrastructure / edited
   by William Lehr.
      p.   cm.
   Includes bibliographical references and index.
   ISBN 0-8058-1610-0 (hard cover)
   1. Telecommunication systems—United States—Reliability.
   2. Telecommunication—United States. I. Lehr, William.
   TK5102.3.U6Q33   1995
   384′.0973—dc20   94-38853
                                                               CIP

Books published by Lawrence Erlbaum Associates are printed on acid-free
paper, and their bindings are chosen for strength and durability.

Printed in the United States of America
10 9 8 7 6 5 4 3 2 1

Contents

PART II: REGULATORY PRACTICE

PART III: EMPIRICAL TRENDS AND EVIDENCE

Introduction

William Lehr
Columbia University

Most of us take the high quality and reliability of telecommunications infrastructure for granted. When we pick up the phone, we just expect it to work.
Following the California earthquakes in 1989 and 1994, telephone service remained operational even when other basic utilities were interrupted. The performance standards for modern networks keep increasing.[1] Meanwhile, the
deployment of digital technology, high capacity fiber optics, and software control
promise a wealth of new services ranging from relatively mundane enhanced
calling options such as voice mail, call forwarding, and automatic number identification to more advanced services such as virtual private networks, integrated
multimedia, and value-added information services. Few of the products or services
we experience in our daily lives have offered such a history of steadily improving
quality and reliability. However, few services are as critical to the smooth operation of modern economies as a communications infrastructure.

Around the globe governments are reexamining their information infrastructure policies.[2] There are two polar views of what the best policy is. At one
extreme lie the proponents of strong government intervention. They view the
information infrastructure as an essential public utility which, if not owned by
the government, should at least be subject to strong government control. At
the other extreme are those who argue that free markets do the best job of
allocating investments, deciding what products to offer, and setting prices. Historically, the proponents of strong government regulation have prevailed, but
current trends favor free market solutions. Because the United States has led
the global movement toward deregulation, perhaps it is logical that the United

1

States lead the way in confronting the issue of infrastructure quality posed by the movement toward reduced regulation and increased competition.

The birth of this volume rests with a policy debate that began to take shape in the United States in the late 1980s when a series of well-publicized incidents raised public concern that the reliability of the telecommunications infrastructure was threatened. The first of these occurred in May 1988, when a fire in the Illinois Bell Telephone Company's Hinsdale switching office interrupted service to hundreds of thousands of Chicago area customers for up to a month.[3] In January 1990, a software problem caused AT&T's long-distance network to fail, seriously disrupting long distance calling in the United States for 9 hours.[4] Other notable incidents that occurred in June, July, and September 1991 helped prompt the U.S. Congress into severely criticizing the Federal Communications Commission (FCC) for not being more proactive in protecting our communications infrastructure.[5] At the same time, state public service commissions were initiating proceedings to determine how regulatory policy should be modified to account for changes in service quality.[6]

Is all this attention warranted? What is the appropriate level of quality? Too high a quality, although unlikely to be as undesirable as too low, could result in excessive network investments and prices. How should we define and measure quality when the range of services and our reliance on those services have increased dramatically? Who should or is likely to bear the costs of increased (or decreased) reliability? Should we rely on the market's *invisible hand* to assure the appropriate quality for our communications infrastructure?

This volume offers the first collection of economic and policy research to focus specifically on the quality and reliability issues posed by modern communications infrastructure. Its goal is to introduce new analytical tools and present recent empirical evidence to better inform the public debate regarding regulatory policy.

The engineers and scientists who build telecommunication, computer, cable television, and other types of networks have always needed to confront reliability and quality issues. They have produced a large theoretical and practical literature exploring the challenges of designing, forecasting, monitoring, and controlling network performance. Capacity planning in the face of stochastic demand is extremely complex. They have developed quite complex operations research techniques to minimize the "costs" of building communications networks; however, these models take prices as a given. The engineering cost studies, although crucial for a complete economic analysis, ignore the role of market dynamics or strategic behavior induced by regulatory policy or increased competition. It is these latter concerns that are of central significance to the authors included here.

In light of the dramatic technical, regulatory, and environmental changes that have transformed and are continuing to transform our communications infrastructure, we need to address these economic issues more seriously. We are moving toward a broadband world of interconnected, hybrid public–private networks wherein the traditional boundaries separating computers, telecommunications, and media entertainment will become increasingly irrelevant. In the United States, alliances between media companies and telephone companies, telephone companies and cable companies, or computer companies and telephone equipment manufacturers are creating new types of information service providers. Local exchange carriers want to enter into long distance transmission, whereas long distance companies are investing in local access facilities. Cable television companies hope to offer telephone services, whereas telephone companies plan to deliver television services. Both types of carriers are exploring more active roles in content provision either as programmers or as vendors of interactive multimedia. Similar alliances are redefining market structures overseas.

These changes clearly demonstrate that we are far from the old world of a single company offering a single service over a single network. Given the large sunk costs associated with constructing a communications network, it seems unlikely that all of these ventures will survive or that the industry that eventually emerges will resemble the economist's ideal of perfect competition. If competition is as intense as many regulators and customers seem to hope, the pressure on companies to cut costs will be extreme. Will these companies sacrifice investments in quality and reliability in the face of aggressive price competition? Or, will the companies be driven to overinvest in quality in order to differentiate their products to protect market shares?

As the industry structure changes, so will incentives to invest in infrastructure enhancements. In a world of interconnected private–public networks, engineers in competing companies will need to coordinate their design efforts. The performance of the whole will depend on the performance of the parts. When control over the infrastructure is decentralized, so is responsibility over guaranteeing system quality and reliability. In addition to agreeing on technical compatibility standards, the owners of the interconnected networks will need to agree on mechanisms for setting prices and exchanging traffic if our hybrid infrastructure is to work efficiently.

Perhaps we should be happy that most of the engineers who design our communications networks have not studied economics, especially game theory. However, engineers are not the only participants whose decisions and opinions will influence the policy debate. Prior to the Congressional report "Asleep at the Switch?" in December 1991, the FCC's interest in reliability and quality was limited.[7] Why fix something that did not appear to be broken? Network capabilities had been improving yearly. Besides, who should know better how to gauge what customers want and the least costly means for meeting those demands than the service providers? If it had not been for calls from constituents and the media attention lavished on the aforementioned incidents that made the reliability of our telephone service suspect, it is doubtful that Congress would have become concerned. Now that public attention has been aroused, we face the danger that we will fix something that really was *not* broken and will *fix it* incorrectly. Students of political and regulatory processes know that although market solutions are often imperfect, they may be superior to even more imperfect regulatory alternatives.[8] Before we can debate our options, however, we need to better understand the alternatives we will face in the future. Enhancing this understanding is the goal of this volume.

The research included here evolved from work that was first presented at a conference organized by the Columbia Institute of Tele-Information in April 1993.[9] The conference brought together a diverse group of academics, government policymakers, and industry executives interested in discussing how changes in industry structure, technology, and regulatory practice were likely to affect the quality of the Public Switched Telecommunications Network (PSTN). Although most of the discussion focused on U.S. telephony experience, the problems and analyses are relevant across the broad spectrum of information technology industries and around the world.

Each of the succeeding chapters presents an essay that addresses a different aspect of the policy debate. The chapters are organized loosely into three sections addressing economic theory, regulatory practice, and empirical evidence. The selection of essays is designed to show the breadth of issues. Each can stand on its own as an original research contribution. Together, the chapters suggest a number of important conclusions that should inspire us toward further research.

Before discussing the authors' contributions, it is worthwhile distinguishing the usual interpretations of the terms *reliability* and *quality* as used in the policy debate and how these terms are meant to be interpreted here. Traditionally, when regulators and telephone companies have discussed *quality of service* they have been referring to a relatively small set of performance measures associated with basic telephony services. These have typically included measures of such things as customer satisfaction levels, dial tone delay, transmission quality, on-time service orders and equipment failure, or call blocking (events that result in call failure).[10] The term *reliability* has been used to refer to the likelihood that network facilities will be operating and available.

This book interprets quality more broadly. It includes the traditional definitions and much more. Quality can refer to any nonprice product or network feature that either enhances customers' perceived value for the services that are provided by the network or lowers the cost of providing the service. Reliability, in all of the sundry ways engineers define and measure it, is just another dimension of quality. Increasing the variety of products offered can be viewed as enhancing quality if customers value additional variety. Technological advances that either lower the costs of existing services or offer improved performance at the same cost also enhance quality. Economists assume, however, that improving quality (either of the products or of the productive facilities) involves net investment and, hence, is costly.[11]

Although the traditional distinction between quality of service and reliability is too limited within the context of the present volume, the traditional distinction captured some important differences. Assuring reliability is very different from improving service quality. The latter is observable in a statistical sense, at least in principle. By collecting data on millions of calls, we can track trends in service quality. Reliability, on the other hand, is much more difficult to predict. By design, the failures by which we define system reliability are rare. We hope that events such as a nuclear power plant disaster are so rare as to (almost) never happen. This difference helps explain why policy has addressed these two issues separately. For example, to enhance reliability, it may be useful for carriers to cooperate closely in their facilities planning and to interconnect so that if one fails the other can provide backup.[12] On the other hand, if we believe competition will encourage carriers to improve their service quality, too much cooperation may be harmful. This volume adheres to the broader, more abstract definition of quality in order to expand our collective thinking about what is and/or should be relevant.

Part I presents four different approaches to the economic modeling of issues that are relevant to this debate. Mainstream economic theory is ill equipped to address the effects on incentives and behavior when traditional models of regulation or imperfect competition are modified to include the choice of quality/reliability as a strategic variable, especially within the context of network industries. The four chapters of Part I present new work that attempts to expand the frontiers of our knowledge.

In chapter 1, Economides and Lehr (E&L) develop a model of imperfect quality competition wherein firms produce complex, interconnected systems. They examine the effects of altering the ownership structure of network facilities on carriers' incentives to invest in improving system quality. Their results suggest that when the ownership of network components is decentralized, strategic incentives to improve quality are reduced. This analysis implies that the divestiture of AT&T and policies that prevent the reintegration of long distance and local access services affect total welfare and network quality adversely. Competition by nonintegrated carriers does not help.

The best free-market outcome analyzed by E&L occurs with competition between two or more vertically integrated carriers, but even this outcome falls far short of the socially optimal solution. Thus, some form of quality regulation may be beneficial even if integrated end-to-end competition proves sustainable.

In chapter 2, Chakravorti and Spiegel (C&S) present a quite different model of carrier behavior and the effects of competitive entry on network quality. A regulated carrier, who is initially a monopolist in both the residential and commercial markets, is facing potential competition in the latter. Increased competition in the commercial market improves quality and consumer surplus for commercial customers but offers less potential for cross-subsidies to the residential market.[13] C&S consider the problem faced by the regulator seeking to balance the trade-off between higher quality for commercial customers and cross-subsidies for residential customers. In order to achieve the goal of universal access, policymakers have felt it was necessary to cross-subsidize access services for low-income customers or expensive-to-serve rural customers. C&S's analysis highlights the conflict between these politically sensitive goals and increased competition. This trade-off will become increasingly important as the transition toward full deregulation progresses.

In chapter 3, Stolleman offers yet another approach to modeling carrier behavior. He analyzes the investment, pricing, and output decisions of a regulated, local exchange carrier who also participates in unregulated, enhanced-services markets. Under the assumption of Open Network Architecture (ONA) restrictions, the carrier is precluded from price discriminating for the basic services. These are sold both in regulated markets and as a platform for enhanced services. His framework provides a mechanism for evaluating alternative regulatory schemes and their effects on incentives to invest in quality enhancements. He focuses on the distortions induced by alternative regulatory frameworks and finds that incentive-based price caps are superior to traditional rate-of-return regulation. In addition, he identifies the importance of adjusting price cap formulae to account for investments that enhance the quality of network facilities to avoid risking that these investments will be deferred. Stolleman cautions against price cap adjustments that are overly sensitive to fluctuations in overall market demand.

Each of the first three chapters approaches the quality issue differently. The models of C&S and Stolleman focus explicitly on the behavior of a regulator; whereas E&L are more interested in competitive dynamics among unregulated firms. All three parameterize quality with a unitary index to simplify the analysis. This approach is common in the literature and with some additional mathematical sophistication proves less restrictive than on first impression (which is important because most people do not think of the quality of a product as being defined simply in terms of a simple ranking of alternatives). For example, telephone quality may be measured in a variety of dimensions such as line noise, circuit availability, bit error rates, customer service, and so on. In certain cases, one could reinterpret the quality index as a vector of indices that would complicate the analysis without significantly altering the results. Or, under suitable, yet strong, assumptions regarding the nature of consumer preferences and cost relationships, multidimensional tastes may be mapped into a unitary index. Finally, we may choose to view the models as presenting partial equilibrium results that focus on a single dimension of quality improvement.

The models also differ in their focus on how quality enters as a strategic variable. E&L and C&S interpret quality as influencing consumer demand (i.e., consumers prefer higher quality at the same price), whereas Stolleman interprets quality as influencing the productivity of network capital (i.e., higher quality networks have lower operating costs). All of the models, however, focus on equilibrium outcomes and so can say little about the actual time it may take to approach predicted outcomes. These models are most properly interpreted as abstractions that are designed to make explicit the logic or *illogic* of arguments, such as "Competition will improve quality" or "Deregulation will lower prices without affecting quality," and to highlight the underlying questions and assumptions that require further study.

The last chapter in Part I, chapter 4, by Chakravorti, Sharkey, and Srinagesh (CS&S), provides a survey of recent pricing theory as it relates to broadband networks. In addition to providing one of the best overviews of traditional demand versus cost-based pricing that is available in the literature, CS&S describe new work that addresses the challenges of sustainable, subsidy-free pricing when a common platform is used to deliver multiple services. CS&S describe a model that allows one to compute the cost-minimizing design parameters for multiple services that have different tolerances for delay or blocking probability. The lower the delay or blocking probability, the higher the service quality. CS&S compute subsidy-free prices for the optimal, quality-differentiated product line that are sustainable in the face of competition. Given that most analysts believe that video services will be susceptible to declining average costs, price discrimination may be essential if carriers are to recover their costs. The design of quality-differentiated product lines may play an important role in implementing such pricing strategies. Whereas the authors of the first three chapters assumed relatively simple, uniform pricing, CS&S discuss more complex, nonlinear strategies.

Part II focuses more directly on the actual regulatory process. Chapter 5 provides a useful bridge between the theoretical research sampled in Part I and its application to policy formulation. In chapter 5, Berg provides a case history of his experiences in attempting to modify the regulatory process with academic inputs. On behalf of the Florida Public Utility Commission (PUC), Berg first developed a theoretical model that demonstrated the superiority of a weighted quality-of-service index for assessing the performance of local exchange carriers. In the past, the PUC relied on a set of pass–fail minimum performance standards for a large number of quality-of-service measures. According to Berg, the justification behind many of these standards is suspect, and there was no systematic basis for comparing different measures. The carrier had little incentive to do more than meet the minimum standard, and meeting these minimum standards could result in resources being allocated inefficiently. For example, the carrier might be induced to expend resources meeting a minimum quality standard for an unimportant measure rather than devoting those resources to exceeding the standard for a more important one.[14]

Berg recommended creating a weighted index that would reflect the regulators' or consumers' preferences regarding trade-offs among the multiple quality measures (i.e., in effect doing in practice what the models in the first three chapters assume implicitly). This unitary index could be used to establish targets and performance incentives that, by allowing the carrier flexibility to allocate resources efficiently, would encourage higher quality service overall at a lower total cost. Concurrently with trying to publish his theoretical contribution in the academic press, Berg attempted to help the Florida PUC implement his proposal. Although the Florida PUC eventually did implement his index, they chose to retain the multiple minimum quality standards, thus eliminating much of the potential gain promised by the new regulatory approach. Berg's experiences provide sobering evidence of the difficulties of moving from economic theory to actual policy.

In chapter 6, Lawton proposes a set of broad guidelines or principals with which to frame the goals for carrier regulation and performance evaluation. Rather than propose a specific model or make definite recommendations regarding the need to either encourage greater or lesser investment in quality enhancements, Lawton seeks to establish a framework for assessing economic performance in the emerging world of hybrid public–private networks. His approach is quite flexible, which reflects the real-world perspective of a researcher who has been active in the formulation of state regulatory policy for many years. He stresses the importance of recognizing that consumer preferences for investments in higher quality networks are likely to be heterogeneous, and regulators ought to consider these tastes in their cost allocation decisions.[15]

Hazlett's contribution in chapter 7 widens the focus of this volume by addressing the response of the cable TV industry to changing regulatory policy. Policymakers have been kind to economists, if not consumers, by offering us a chance to observe several regime shifts: from regulated to deregulated to regulated again. The examples he considers are cable rate deregulation in California in 1979, Federal deregulation in 1984, and Federal re-regulation in 1992. Hazlett argues that cable operators respond to increased (decreased) regulation by reducing (increasing) service quality. According to Hazlett, this is exactly what one should expect to see based on standard microeconomic theory. If regulators constrain the nominal price below the free-market level, then the firms will reduce their investments in quality-improving inputs, which lowers demand so that a new equilibrium is possible at the lower (regulated) price.

Hazlett's analysis illustrates the complex mechanisms by which firms adjust their service quality. It is easier to observe such strategic behavior among cable firms both because of the happenstance of regulatory regime shifts and because cable operators may more easily and quickly adjust their service quality in response to exogenous forces such as regulation. With both cable TV and telecommunications networks, it is relatively difficult to adjust the quality of sunk investments in network capacity, although prospectively, future enhancements may be adversely affected (as argued by Stolleman in chapter 3). However, it is relatively easy to reduce variable expenditures on programming and service maintenance. As we move toward integrated broadband networks, the adverse effects of regulation on service content quality will become more important. Moreover, regulating content quality is significantly more difficult than regulating the technical parameters of service quality. In addition to being more highly subjective, content restrictions may run afoul of free speech rights (e.g., the First Amendment in the United States).

Hazlett supports his arguments by analyzing market performance before and after regulatory regime shifts and finds that following deregulation, cable penetration increased both in absolute terms and in terms of market share vis-à-vis over-the-air broadcast services. Moreover, spending on programming and investment in capacity expansion increased. Both trends suggest an improvement in quality following deregulation. In contrast, following re-regulation, initial evidence suggests lower investments in programming and retiering so as to reduce the quality of basic-tier cable television services.

In addition, Hazlett examines the political support for and against the Cable (Re-)Regulation Act of 1992 and finds that the alignment of interests coincides with an argument that the expected effect would be an *increase* in quality-adjusted prices, which is exactly the opposite of the avowed purpose of the Act. In sum, Hazlett's contribution represents a broad condemnation of active cable regulation and its adverse implications for quality, which is illuminating for regulators everywhere confronting the need to regulate or de-regulate public information infrastructure.

Part III includes three chapters that offer differing perspectives on the current and prospective state of our telecommunications infrastructure. In chapter 8, Kraushaar describes recent data collection efforts by the FCC to expand its tracking of quality-of-service and network reliability data. Today, the FCC collects statistics on a greatly expanded array of service quality measures. Moreover, these data are collected and made available to the public electronically via a public bulletin board. The data collection effort described by Kraushaar is part of the FCC's ARMIS system, which represents a larger effort to improve the *quality* and comparability of data on telecommunications carriers. The ARMIS system will help reduce the costs of state regulation and may provide the basis for more systematic benchmarking.

Because the data series is quite new, comparisons with historical trends are impossible in all but a few cases. Happily, those measures that admit to limited trend analysis appear to indicate that service quality has improved slightly. For the remainder of the chapter, Kraushaar discusses the characteristics of the data, highlighting some of the problems with making intercompany comparisons because of differences in reporting standards. The present effort focuses on collecting raw measures rather than company indices. This should allow the states greater freedom to construct their own indices and may make it more difficult for the companies to obscure performance problems. Hopefully, the new data will enable us to establish a benchmark against which future trends may be compared.

In chapter 9, Tomlinson analyzes the performance of Competitive Access Providers (CAPs) such as Teleport and Metropolitan Fiber Systems and their impact on the Bell Operating Companies (BOCs). Although direct comparisons are difficult, his survey suggests that the CAPs have offered superior reliability. He credits the CAPs initial clientele for encouraging this high quality. The first CAP clients were either long distance telephone companies (IXCs) seeking connections among their Points of Presence (POPs) or large corporate customers who wished to connect directly to an IXC, bypassing the BOC's local access network. These sophisticated customers demanded extremely high reliability services for their backbone connections. Tomlinson believes that the net effect of this competition, at least initially, has been to spur the BOCs to enhance their service quality and lower their prices. The combination of these two trends bodes well for the customers of these special access services (although as suggested by the model of C&S in chapter 2, the effect on residential consumers may be negative as long as they are served by a regulated monopolist).

The higher reliability is feasible because of the new technology that is being deployed by the CAPs. This includes self-healing fiber ring topologies, redundant electronics, and integrated network management software. Furthermore, the market opportunity for the CAPs is partially due to the increasing demand from customers for highly reliable services. Businesses that are critically dependent on the communications infrastructure seek the route diversity achieved when services are obtained from multiple carriers. If the BOC network fails, then the CAP facilities can handle the customer's traffic. This assumes, however, that the two networks do not fail simultaneously, which they may if they are closely enough coupled and/or share the same conduit where the failure occurs.[16] Although the initial effects of increased competition appear promising, coordination and cooperation will become increasingly important as we evolve toward a network of interconnected networks and from simple dedicated transport to more complex, switched services offerings.

In chapter 10, Wohlstetter considers the implications for reliability of software-controlled networks. The same software control that offers us a plethora of new services and the potential for anticipating and correcting congestion problems automatically may present an achilles heel for network reliability. In the old world of Ma Bell, there was a single telephone company that owned everything from hand set to hand set. There were many single points in which the network could fail, but the effects of these failures tended to be localized. With software control, a single programming bug can crash the whole network. In addition, as we unbundle network services through such policies as Open Network Architecture (ONA) and customers are permitted to access network control functions directly, the network is more vulnerable to malicious attack and/or mistakes by nontelephone employees resulting in major network failures. According to Wohlstetter, increased vulnerability to catastrophic failure is one of the collateral costs of increased openness.

Wohlstetter offers a lively account of recent experiences with software bugs, hackers, and other types of network threats that were not relevant a decade ago. He questions our ability to protect against these threats given the current state of regulations and liability law. Wohlstetter calls for software "fire walls" that would protect the core of the network from customer mistakes or attacks and "cyber-crimes" to enable better prosecution of malicious attackers and "tele-torts" to facilitate the pursuit of civil remedies. He suggests that we may need a set of rules for software like those that were used to certify hardware equipment connected to the PSTN (i.e., the Part 68 Rules). Wohlstetter argues that we need to have responsibility for network reliability follow control, which means that as customer control over system-level software increases so should their responsibility for reliability.

The 10 chapters in this book introduce a diverse set of perspectives that need to be included in the formulation of future policy. They also suggest some important gaps in the literature that require additional research. Three of these are worth special attention. First, all of the chapters in Part I assume that quality is perfectly observable and that the carrier is perfectly informed about its costs and the nature of consumer demand.[17] This is an unrealistic assumption that ignores important information problems. It may be costly for customers to discern service quality, especially when different portions of the network are provided by different firms. There is a rich economics literature that addresses the problem of imperfect information in product markets.[18] Firms alter their behavior to establish positive reputations or otherwise signal to customers the high quality of their products. This literature has not yet been extended to the case of network industries.

The problem of network reliability is more difficult to analyze because the asymmetric information problems are compounded by real uncertainty. The carriers may not know the true reliability of their networks and be reticent to share what information they have with potential competitors. How do we assess the overall reliability of the interconnected infrastructure? Models of insurance markets and financial stability may be more appropriate for evaluating these types of information problems.

The political economy of regulatory processes vis-à-vis quality is another area that merits additional study. In order to evaluate nonmarket solutions to a perceived quality problem, it is essential to evaluate the behavior of the regulatory institutions. Once again, a large economics literature has examined how procedures and institutional structure influence regulatory outcomes (e.g., what information is collected, who is allowed to participate, the timing of review cycles, etc.).[19] For example, if policymakers believe they will be unduly penalized for a major network failure, they may propose excessively restrictive minimum quality standards. Even if meeting these leads to much higher costs in aggregate, the cost increase to each customer may be sufficiently small so that no one complains. In this situation, policymakers would be overly risk averse. Berg's experience suggests that this may be the case in Florida.

Alternatively, policymakers may fail to approve necessary quality-enhancing investments in order to keep prices low, believing that should reliability problems arise, others will bear the blame. In this case, the regulatory process would discourage adequate investments in quality. Additional research is needed to distinguish between these two cases and to determine whether it might not be better to limit regulatory interventions in this area.

Finally, we need to continue to collect and monitor quality or reliability performance data. The work presented in Part III provides a useful start, but it is still too early and the data series are too incomplete to permit an accurate assessment of trends. The mere fact that the carriers know that policymakers are watching may provide positive incentives. If states differ in how they choose to implement incentives to enhance quality, then a comparison of their experiences would be useful in identifying good policies.

Data are particularly scarce when it comes to assessing the costs and benefits of improved reliability. Businessmen often talk as if they have lexicographic preferences such that they would always prefer a more reliable telecommunications service, regardless of the cost differential. Are these businessmen excessively risk averse? Do firms over- or underinvest in redundant systems? To answer these questions we need a better understanding of the costs of network failures to both the carrier and to the customers.

The chapters collected here provide a useful starting point for what I hope will be an ongoing research effort. We should be warily encouraged that the limited empirical evidence in Part III suggests that competition and deregulation do not appear to have resulted in significant deterioration in the quality of our communications infrastructure; and the contributions in Parts I and II support the idea that, properly managed, deregulation and increased competition can assist in realizing additional improvements in the future. However, the challenges we face are complex and these preliminary assessments warrant on-going review.

ACKNOWLEDGMENTS

In addition to the dedicated efforts of the authors, this volume benefited greatly from the comments, helpful suggestions, and administrative support provided by Eli Noam, Gary Ozanich, Michelle Wilsey, and the staff of the Columbia Institute of Tele-Information. I would especially like to thank Annie Kang, who provided invaluable assistance in organizing the April 1993 conference at Columbia University where this research was first presented. The following individuals who served as discussants at the conference provided valuable comments that improved the quality of the final product, even when their opinions differed significantly from those expressed herein: Jerry Brockhurst, Jill Butler, Art Deacon, Gerald Faulhaber, Nelson Ledbetter, Greg Lipscomb, Page Montgomery, Al Novell, David Peyton, Ed Regan, Leonard Sawicki, Sid Shelton, Richard Steinberg, and Steven Wildman. Finally, I would like to thank the series editor, Chris Sterling, and my editor at Lawrence Erlbaum Associates, Hollis Heimbouch, for helping to assure the successful completion of this project.

ENDNOTES

1. For example, in chapter 9, Tomlinson reports that the Competitive Access Providers are supporting average circuit availability of 99.999%, which translates to a downtime of less than 5.26 minutes per year.

2. For a survey of trends in regulatory policy, see *New Directions in Telecommunications Policy*, edited by Paula Newberg (Durham, NC: Duke University Press, 1989) or *Telecommunications in Europe*, by Eli Noam (New York: Oxford University Press, 1992).

3. See J. Wiggins and J. Wiggins, "Economic Consequences of the Hinsdale, Illinois Bell Fire of May 8, 1988," memorandum prepared for National Science Foundation under grant #CES-8820941 by Crisis Management Corporation, Redondo Beach, CA, September 1989. They reported that the fire left "38,000 people, including over 3,500 businesses, without telephone and data communication links and 475,000 more customers with limited service" and that "restoring service to full capacity took over a month" (p. 1).

4. See U.S. Congress, "Asleep at the Switch? Federal Communications Commission Efforts to Assure Reliability of the Public Telephone Network," Fourteenth Report by the Committee on Government Operations, House Report 102-420, Government Printing Office, December 11, 1991, p. 2.

5. *Ibid.* In June and July 1991, there were a series of incidents affecting millions of customers of Bell Atlantic and Pacific Bell (p. 2), and in September 1991, "a power failure at an AT&T switching facility in lower Manhattan cut off most long distance telephone communications to and from New York City as well as air traffic to the area's three major airports" (p. 3). The report's first three findings were as follows:

    1. The public switched networks are increasingly vulnerable to failure; and the consequences for consumers and businesses and for human safety are devastating.

    2. The problem of network reliability will become increasingly acute as the telecommunications market becomes more competitive.

    3. No Federal agency or industry organization is taking the steps necessary to ensure the reliability of the U.S. telecommunications network. (p. 4)

6. See chapters 3, 5, and 6 by Neal Stolleman, Sanford Berg, and Ray Lawton, respectively, for a diverse set of views on some of the issues raised by attempting to incorporate quality measures into regulatory policy.

7. See Kraushaar's contribution for a discussion of the FCC's efforts with respect to quality and reliability.

8. See Berg's contribution for a discussion of some of the problems faced in the regulatory process.

9. The Columbia Institute of Tele-Information (CITI), part of Columbia University's Graduate School of Business, is a research center devoted to the independent study of economic, policy, and management issues involving telecommunication, computer, and electronic mass media.

10. See, for example, the Organization for Economic Cooperation and Development (OECD) report #22, *Performance Indicators for Public Telecommunications Operators* (Paris: OECD, 1990).

11. According to followers of Professor Deming's philosophy of Total Quality Management (TQM), it is often possible to simultaneously improve customer value and reduce costs. TQM is a managerial process for approaching what the economist refers to as the efficient production frontier. Economists focus on characterizing this production frontier, implicitly assuming that the firm is well managed, which may mean that the firm is employing TQM strategies.

12. Close coupling of carrier networks may actually decrease reliability if a failure on one network can produce the failure of the other network. Additional research into the effects of backup supply agreements is necessary before we can be confident that these will improve reliability.

13. C&S assume that the competing commercial carrier is fully integrated, so their result of improved quality coincides with the result of E&L, even though the two modeling approaches differ significantly.

14. In economic terms, the net social benefit associated with exceeding one standard may exceed the net social benefit associated with meeting another standard.

15. The models presented in chapters 1, 2, and 4 explicitly address the problems posed by this heterogeneity.

16. In some cases, the CAP leases BOC facilities to offer its services so the routing diversity may be more perceived than real.
17. Implicitly, regulators are assumed either to be imperfectly informed about firms' costs or otherwise unable simply to compute and enforce the socially optimal outcome.
18. See, for example, Stiglitz, J., "Chapter 13: Imperfect Competition in the Product Market," in R. Schmalensee and R. Willig (Eds.), *Handbook of Industrial Organization*, New York: North-Holland, 1989.
19. See, for example, Noll, R., "Chapter 22: Economic Perspectives on the Politics of Regulation," in R. Schmalensee and R. Willig (Eds.), *Handbook of Industrial Organization*, New York: North-Holland, 1989.

Part I

ECONOMIC THEORY OF NETWORK QUALITY

Chapter 1

The Quality of Complex Systems and Industry Structure

Nicholas Economides
New York University

William Lehr
Columbia University

1. INTRODUCTION

Most products may be thought of as complex systems, or **networks**, composed of subsystems and components. This is especially true of many information technology products. Networks may be physical, as in the case of the telephone, cable television, or Internet communication networks; or they may be logical, as in the case of the different software modules/layers that must interact to support an application such as word processing, electronic mail, or customer billing.[1] The quality of these networks depends on the quality of the constituent subsystems. For example, the clarity of a long distance call depends on the qualities of the telephone sets on both ends, the originating and terminating local exchange networks, and the long distance carrier's network. Coordinating design and investment decisions to assure appropriate quality-of-service levels is difficult enough when all of the components are owned by the same company. What happens when different firms own different parts of the network? Should we be concerned that changes in industry structure will lead to reductions in the quality of our information infrastructure?

To answer these questions, we develop a model of quality competition with complementary components. Each firm produces either or both of two components that must be combined to create a usable system. For example, the components may include local access and long distance, computer hardware and software, or a stereo receiver and speakers. We examine how changes in industry structure affect firms' pricing and product-design behavior. Although subject to

17

important caveats, this analysis yields four results that should be of interest to policymakers.

First, *network quality and total surplus are higher and prices are lower when there exists a vertically integrated firm offering a complete system.* This suggests that the reintegration of local and long distance carriers may improve total welfare and incentives to invest in higher quality telecommunications infrastructure.

Second, *effective competition is not sustainable in the face of a bottleneck facility, even if interconnection is required, unless regulations also control the price of access to the bottleneck.* The introduction of quality competition provides another strategic variable that can be manipulated by the owner of the bottleneck facility to foreclose downstream competition.[2]

Third, *competition among integrated producers improves total welfare, leading to lower prices, a larger total market, and increased quality or variety available.* This suggests that there will be a social benefit from having multiple *integrated* carriers offering long distance, cable television, and other information technology services.[3]

Fourth, *once we have two integrated producers competing, there are no gains from requiring interconnection.* The firms will behave so as to deny demand to hybrid system markets (i.e., systems that are composed of components from two or more firms). Thus, introducing quality as a decision variable reverses the results of Matutes and Regibeau (1988, 1992) and Economides (1988, 1989, 1991) in which vertically integrated producers will choose to interconnect their subnetworks so as to offer hybrid systems.

Four features should be considered when interpreting the results of this chapter. First, we assume consumers differ only with respect to their willingness to pay for improvements in quality beyond a minimum quality level. Consumers rank all products identically according to a unidimensional quality index.[4]

Second, if firms choose to produce identical quality products, then these products will be perceived by consumers as perfect substitutes, and competition will result in marginal cost pricing. To avoid this, firms will choose to quality-differentiate if they choose to offer products of higher than the minimum quality.

Third, firms cannot price discriminate among consumers except by manipulating component versus system prices (i.e., product bundling) and by offering multiple quality levels. This precludes complex price discrimination strategies such as two-part pricing or volume discounting (e.g., WATs services).

Fourth, we ignore positive demand externalities (network externalities) that do not arise as part of the particular mix-and-match structure that we examine.[5] For example, a telecommunications network exhibits network externalities if consumers typically are willing to pay more to join a larger network because they have more options of who to call; similarly, consumers may prefer to purchase the more popular hardware platform because more software is available. Our model excludes network externalities that arise **outside** the model. However, interconnection and compatibility that are analyzed in the model may increase the demand of some components, and this may be considered a network externality. The effect on quality equilibria of including positive externalities that arise outside our model is ambiguous. It could either increase or decrease network quality, depending on how customer markets are served. Addressing this issue is an important topic for future research.

The rest of this chapter is organized into four sections. Section 2 describes our model, its relationship to other models in the literature, and our solution approach. We proceed by analyzing how changes in industry structure alter equilibrium solutions in a number of cases. These cases were chosen both to isolate the effects of specific changes in the regulatory and/or market environment and to correspond to real-world situations in which network quality is an issue. The cases are distinguished by (a) the number of firms, (b) the degree of vertical integration of each of the firms, (c) whether firms are allowed to price discriminate, and (d) whether integrated firms' components are compatible (i.e., are hybrid systems available?). Section 3 derives the results by applying the approach outlined in section 2 to each of the cases summarized in Fig. 1.1. Readers who are more interested in the interpretation of these results may skip this section. Section 4 interprets our cases in light of the real world, summarizes our four principal conclusions, and suggests opportunities for future research.

2. THE MODEL SETUP

There is a rich literature that analyzes imperfect competition when firms use quality to differentiate their products.[6] Traditional models focus on a market for a single product that is available at different quality levels. There is a continuum of consumer types who rank the products identically but differ in their willingness to pay for quality. For example, a representative consumer of type θ who buys one unit of a product of quality q at price p receives utility:

$$U_\theta(q, p) = \theta q - p, \tag{1}$$

or, if the consumer chooses not to purchase a good, he or she receives a reference utility U_θ^0, which is normalized to zero. When coupled with an assumption regarding the distribution of consumer types, these preferences allow one to compute the demand for each product quality level. For example, it is common in the literature to assume that consumer types are distributed uniformly on the unit interval, or:

$$\theta \sim \text{Uniformly on } [0, 1]. \tag{2}$$

Then in a market with a single product of quality q, all consumers of types $\theta \geq p/q$ will purchase the product, thus yielding a demand of $1 - (p/q)$.

Section 3.1   Vertically Integrated Monopolist

3.1.1 Single-product vertically integrated monopolist
3.1.2 Multiproduct vertically integrated monopolist

Section 3.2   Bilateral Monopoly: Independent (Non-Integrated) Monopolists Upstream and Downstream

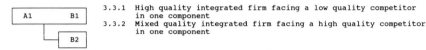

Section 3.3   Duopoly: Integrated Firm Facing Competition in One Component

3.3.1 High quality integrated firm facing a low quality competitor
       in one component
3.3.2 Mixed quality integrated firm facing a high quality competitor
       in one component

Section 3.4   Three Independent Firms, Each Producing a Single Component. High Quality Upstream and High or Low Quality Downstream

Section 3.5   Duopoly Competition Between Two Vertically Integrated Firms

3.5.1 Duopoly Competition of Vertically Integrated Firms With Incompatible
       Components
3.5.2 Duopoly Competition of Vertically Integrated Firms That Produce
       Compatible Components

FIG. 1.1.   Industry structure cases. Numbers identify firm; letter identifies component.

Following Hotelling (1929) and Shaked and Sutton (1982), we model the firms as competing in a two-stage game. In the first stage, the firms choose quality specifications for their products, and in the second stage they choose prices (see Fig. 1.2).[7] In the second stage, price competition leads to zero prices whenever both firms' products are perfect substitutes (i.e., have identical quality levels). Firms that produce products of different quality may charge different prices in equilibrium and divide the market. For example, let one firm offer a high-quality product of quality q_H at price p_H and the other firm offer a low-quality product of quality q_L and price p_L. Then there will be a marginal consumer with type θ_{L0} who is indifferent between purchasing the low-quality product and no product,

$$\theta_{L0} = p_L/q_L \in [0, 1], \tag{3}$$

and another consumer with type θ_{HL} who is indifferent between purchasing the high- or low-quality product,

$$\theta_{HL} = (p_H - p_L)/(q_H - q_L) \in [\theta_{L0}, 1]. \tag{4}$$

Demand for the high-quality product will be $1 - \theta_{HL}$, and demand for the low-quality product will be $\theta_{HL} - \theta_{L0}$.[8] Noncooperative equilibrium price levels $p_1^*(q_1, q_2)$, $p_2^*(q_1, q_2)$ are determined in the second stage game as solutions to $\partial\Pi_1(p_1, p_2, q_1, q_2)/\partial p_1 = \partial\Pi_2(p_1, p_2, q_1, q_2)/\partial p_2 = 0$. Equilibrium quality levels are

In each case an industry structure is defined by

- the number of firms (1, 2, or 3)
- the vertical market structure (each firm produces component A, B, or both)
- whether price discrimination is allowed (different prices for components and bundles allowed?)
- whether components are compatible (are hybrid systems feasible?)

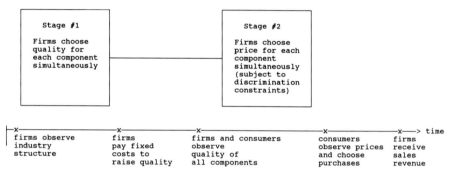

Game assumes:

- Perfect information
- Consumers distributed uniformly on unit interval, $\theta \in [0, 1]$ with $U_\theta(p, q) = \theta q - p$ if purchase; $= 0$ otherwise. p is price of product; q is quality of product $= \min(q_A, q_B)$.
- Costs of quality improvement $= \frac{1}{4}q_i^2$, where q_i is the quality of component $i = A$ or B.

FIG. 1.2. Structure of the quality price game.

found in the first stage of the game as solutions of $d\Pi_1(p_1^*(q_1, q_2), p_2^*(q_1, q_2), q_1, q_2)/dq_1 = d\Pi_2(p_1^*(q_1, q_2), p_2^*(q_1, q_2), q_1, q_2)/dq_2 = 0$.

This framework is useful for analyzing the determination of equilibrium qualities and prices. It can also be used to determine how changes in industry structure (e.g., entry/exit) affect firms' product design (i.e., quality choices) and pricing behavior in equilibrium. We extend this basic framework of vertical product differentiation by assuming that each demanded composite good (or system) consists of a network of two components, A and B, with qualities q_A and q_B. Consumer preferences are defined with respect to the quality of the composite good, q_{AB}, which is equal to the minimum of the component qualities:

$$U_\theta(q_{AB}, p_{AB}) = \theta q_{AB} - p_{AB} \qquad (5)$$

where,

$$q_{AB} = \min(q_A, q_B) \in [0, \infty) \text{ and } U_\theta(0, p_{AB}) = 0 \qquad (6)$$

and p_{AB} is the price that the consumer pays for the composite good or system.

If firms' components are **compatible**, then it is possible to create a **hybrid system** by combining the A component from one firm with the B component from another firm. The price for this system is equal to the sum of the individual component prices. If a firm is allowed to **price discriminate**, then it may charge a price for a bundled system (composed of two components from the same firm), which is different than the sum of the firm's individual component prices.[9] The regulatory policy known as Open Network Architecture (ONA) prohibits this form of price discrimination. ONA requires common carriers to offer component pricing (and hence, interconnection via compatible components), which does not discriminate between hybrid and bundled systems.

We assume that there is an increasing, convex fixed cost associated with improving quality above the minimal level, but that marginal costs are zero.[10] Specifically, we assume that the cost to a firm of providing component i of quality q_i is:

$$C(q_i) = q_i^2/4, \text{ where } i = A, B. \tag{7}$$

A firm's total costs are computed as the sum of the component costs for each component produced by the firm. Thus, *there are scale economies associated with quality but no scope economies from vertical integration.*

A strategy for a firm that produces n components is a choice of qualities for each of the components in the first stage and a set of prices for each of the components in the second stage, conditional on *all* of the firms' quality choices from the first stage. A solution to this game consists of a set of equilibrium quality and pricing decisions for each firm. We focus on subgame perfect Nash equilibria.[11]

For each of the industry structures described in Fig. 1.1, we determine price-quality equilibria. The game structure is pictured in Fig. 1.2. In each case, we begin by specifying (a) the number of firms and the components each produces, (b) whether price discrimination is allowed, and (c) whether the firms' components are compatible (or interconnected). This determines the range of system products that will be available to consumers. The game is solved recursively, starting with the competitive pricing equilibrium in the second stage. In order to obtain a pricing equilibrium when there are multiple products, we need to assume an ordering of component qualities.[12] If a pricing equilibrium does not exist, then the assumed ordering of product qualities is not part of a subgame perfect Nash equilibrium. If we find a pricing equilibrium, then we substitute this into the firms' profit functions and solve for a quality equilibrium for the first stage. A solution to the game, if it exists, is a set of product qualities and prices that can be used to compute the share of the market served, firm profits, and consumer surplus. Because a firm can always choose to set the quality of every component equal to zero, profits should always be weakly positive. We can interpret a firm earning zero profits as inactive in the market. The threat of entry does not constrain the behavior of active firms because these markets are not contestable (Baumol, Panzar, & Willig, 1982).[13] This procedure is followed for each possible ordering of component qualities and for each of the candidate industry structures identified in Fig. 1.1.[14]

3. ANALYSIS OF THE MODEL

This section derives the quality and pricing equilibria for each of the cases summarized in Fig. 1.1. Readers who are not interested in the mathematical derivation of the results may proceed directly to the discussion in Section 4. The results are summarized in Figs. 1.3 and 1.4.

3.1 Vertically Integrated Monopolist

3.1.1 Single Product Monopolist

We first consider a vertically integrated monopolist who produces a single version of each component, A and B, and sells them as a single product AB. Because consumers care only about the quality of the composite good, AB, their willingness to pay depends on the minimum of q_A and q_B. Because there is never a revenue gain associated with increasing a component's quality above the highest quality component with which it might be paired, and because component costs increase with quality, we should expect the highest quality upstream and downstream components to have the same quality in *all* equilibria. Therefore, the monopolist will choose identical component qualities $q = q_A = q_B$.

Section 3.1   **Vertically Integrated Monopolist**

[diagram: A — B]

$p_{AB} = 0.1250$ $q_{AB} = 0.25$ $s_{AB} = 0.500$ $\Pi_1 = 0.03125$ $CS = 0.03125$
- Monopolist prefers not to offer a second component of either type
- Socially optimal solution with single product, uniform pricing

$p_{AB} = 0.094$ $q_{AB} = 0.375$ $s_{AB} = 0.75$ $\Pi_1 = 0.0$ $CS = 0.1055$

Section 3.2   **Bilateral Monopoly: Independent (Non-integrated) Monopolists Upstream and Downstream**

[diagram: A — B]

$p_A = p_B = 0.0741$ $q_A = q_B = 0.222$ $s_{AB} = 0.333$ $\Pi_A = \Pi_B = 0.01235$ $CS = 0.01235$
$p_{AB} = 0.1481$ $\Pi_{A+B} = 0.0247$

Section 3.3   **Duopoly: Integrated Firm Facing Competition in One Component**

[diagram: A1 B1 ; B2]

3.3.1 Integrated high quality (firm 1) vs. non-integrated low quality (firm 2)
- Firm 1 prices A_1 to foreclose firm 2 from market, and acts as monopolist

3.3.2 Mixed quality integrated (firm 1) vs. high quality non-integrated (firm 2)
- Without price discrimination, firm 1 sets $q_{B1} = 0$ and firms behave as bilateral monopolists (section 3.2)
- With price discrimination allowed, there is no equilibrium

Section 3.4   **Three firms: High Quality Upstream and High (Firm 2)/Low (Firm 3) Quality Downstream**

[diagram: A1 B2 ; B3]

- Without price discrimination, firm 1 and firm 2 foreclose firm 3 from market, and act as bilateral monopolists (section 3.2).

- With price discrimination allowed,
$p_{A1B2} = 0.1448$ $q_{A1} = q_{B2} = 0.222$ $s_{A1B2} = 0.337$ $\Pi_1 = 0.01471$ $\Pi_2 = 0.01050$ $CS = 0.0136$
$p_{A1B3} = 0.0114$ $q_{B3} = 0.0209$ $s_{A1B3} = 0.112$ $\Pi_3 = 0.00013$
     $s_{A1B2} + s_{A1B3} = 0.449$ $\Pi_1 + \Pi_2 + \Pi_3 = 0.02534$

Section 3.5   **Duopoly Competition Between Two Vertically Integrated Firms**

[diagram: A1 B1 ; A2 B2]

3.5.1 Duopoly, both integrated firms with incompatible components
$p_{A1B1} = 0.1077$ $q_{A1} = q_{B1} = 0.253$ $s_{A1B1} = 0.525$ $\Pi_1 = 0.0244$ $CS = 0.0431$
$p_{A2B2} = 0.0103$ $q_{A2} = q_{B2} = 0.048$ $s_{A2B2} = 0.262$ $\Pi_2 = 0.0015$
     $s_{A1B1} + s_{A2B2} = 0.787$ $\Pi_1 + \Pi_2 = 0.0259$

3.5.2 Duopoly, both integrated firms with compatible components
- Firms price so as to foreclose hybrid markets and hence solution defaults to duopoly with incompatible components (section 3.5.1)

FIG. 1.3. Summary of results for industry structure cases. Equilibrium values: p_i = price; q_i = quality; s_i = market coverage; Π_i = profits; CS = consumer surplus.

	--- High --- Price Quality	--- Low --- Price Quality
Social optimum with perfect price discrimination	n/a 0.500	n/a n/a
" " uniform pricing	0.094 0.375	n/a n/a
Vertically integrated monopoly	0.125 0.250	n/a n/a
Bilateral Monopoly: non-integrated upstream/downstream	0.148 0.222	n/a n/a
Three non-integrated (w/price discrimination)	0.145 0.222	0.011 0.021
Duopoly, vertically integrated, not compatible	0.108 0.253	0.010 0.048

	Profits	Consumer Surplus	Total Surplus	Market Coverage
Social optimum with perfect price discrimination	n/a	n/a	0.125	100%
" " uniform pricing	0.000	0.106	0.106	75%
Vertically integrated monopoly	0.031	0.031	0.063	53%
Bilateral Monopoly: non-integrated upstream/downstream	0.025	0.012	0.037	36%
Three non-integrated (w/price discrimination)	0.025	0.014	0.039	36%
Duopoly, vertically integrated, not compatible	0.026	0.043	0.069	65%

FIG. 1.4. Summary of results. Industry structures that are not shown do not have equilibrium solutions and default to one of the outcomes above. For example, the case of three firms without price discrimination defaults to the case of bilateral monopoly, and the case of a non-integrated firm competing against a vertically integrated firm defaults either to the bilateral monopoly case or the vertically integrated monopoly case.

Because the monopolist sells a single composite good, price discrimination based on quality is impossible. Hence, the monopolist will quote a single bundled price, p, for the good AB. The marginal consumer, θ_1, who is indifferent between buying or not buying AB, is defined by:

$$\theta_1 q - p = 0 \Leftrightarrow \theta_1 = p/q \tag{8}$$

All consumers of types $\theta \in [\theta_1, 1]$ will purchase the good; thus the monopolist's profits are:

$$\Pi_1(p, q) = p(1 - \theta_1) - 2(1/4)q^2 = p - p^2/q - q^2/2. \tag{9}$$

These profits are maximized when the monopolist sets his or her price at $p^*(q) = q/2$. This implies that at the first stage the monopolist will choose quality q to maximize:

$$\Pi_1(p^*(q), q) = q/4 - q^2/2. \tag{10}$$

The optimal choice for q^* is $1/4$. This implies that $p^* = 1/8$, $\theta_1 = 1/2$, and $\Pi(p^*, q^*) = 1/32$. Consumer surplus in this case is also $1/32$.[15]

3.1.2 Multiproduct Vertically Integrated Monopolist

The previous analysis considered the case of a single-product monopolist. We show in this section that, given our demand and cost assumptions, *the monopolist will not choose to introduce a second-product quality*. To see this, consider the situation in which the monopolist offers a high-quality upstream/downstream product of quality q_1 and price p_1, and a second lower quality downstream component of quality q_2 and price p_2. In this case, demands for the high- and low-quality systems are $1 - \theta_1$ and $\theta_1 - \theta_2$ respectively, where $\theta_1 = (p_1 - p_2)/(q_1 - q_2)$ and $\theta_2 = p_2/q_2$. The firm's profits are:

$$\Pi_1(p, q) = p_1(1 - \theta_1) + p_2(\theta_1 - \theta_2) - q_1^2/2 - q_2^2/4 \tag{11}$$

which implies that the optimal second stage prices are $p_1^* = q_1/2$ and $p_2^* = q_2/2$.[16] At these prices, the monopolist's profits are:

$$\Pi_1(p^*(q), q) = q_1/4 - q_1^2/2 - q_2^2/4. \tag{12}$$

To maximize these profits, the monopolist should set the quality of the low-quality product equal to zero, $q_2 = 0$. Then the optimal quality of the high-quality product is equal to $1/4$, as before. A zero quality product is identical to the outside good. Therefore, the integrated monopolist will choose not to offer a second quality-differentiated product.

3.1.3 Socially Optimal Solution

Before analyzing other industry structures, it is worthwhile computing the socially optimal solution with a single product under a break-even constraint and no price discrimination. This is obtained by maximizing consumer surplus subject to the constraint that the firm recovers its costs. The maximization problem to be solved is as follows:

$$\max_{\theta_1} \left[\int_{\theta_1}^{1} (q\theta - p)d\theta \right] \tag{13a}$$

subject to:

$$\Pi(q, p) = 0 = p(1 - \theta_1) - \frac{1}{2}q^2 \tag{13b}$$

and

$$q\theta_1 - p = 0 \tag{13c}$$

Equation (13a) defines total consumer surplus that is equal to total surplus when firm profits are constrained equal to zero, as in constraint (13b). Constraint (13c) identifies the marginal consumer with type θ_1. Solving (13b) and (13c) for p and q as functions of θ_1 yields $p = 2\theta_1^2(1 - \theta_1)$ and $q = 2\theta_1(1 - \theta_1)$. After substituting for p and q in (13a), the optimal θ_1^* is $1/4$, which implies that $p^* = (6/64) \cong 0.094$, $q^* = (3/8) = 0.375$, and total surplus is 0.1055. This is the outcome that maximizes total surplus if the firm is constrained to uniform pricing and to offering a single product.

If the firm could perfectly price discriminate in order to recover its costs, then the maximization problem which would define the optimal solution would be: Under perfect price discrimination, there is no reason not to serve the entire market and allow all consumers to enjoy the highest quality product. Therefore,

$$\max_{q} \, [\int_0^1 (q\theta) \, d\theta - \frac{1}{2} q^2] \tag{14}$$

the marginal consumer θ_1^{**} will have type zero, will be charged a zero price; there would be no reason to offer a second, lower quality product.[17] Higher type consumers will pay higher prices. Total surplus is maximized at $q^{**} = 1/4$; this results in a total surplus of $(1/8)$. In principle, this could be distributed between the firm and consumers via lump sum transfers according to whatever allocation seemed desirable.

A comparison of this solution and the single-product/uniform-pricing solution highlights the important role that quality-differentiated product lines may play in implementing second-degree price discrimination. It is common for a firm to offer multiple products that are designed and priced in order to induce customers to self-select so that consumers with a high willingness to pay choose to purchase more expensive products.[18] There is a large literature discussing the design of optimal tariffs to implement this type of second-degree price discrimination that may be based on quality differentiation, volume of purchases, or some other observable attribute that allows customers to be segregated into self-selected groups. Mitchell and Vogelsang (1991) offer an introduction to this literature and its application to telecommunications.

These two solutions provide benchmarks against which to compare the outcomes under alternative industry structures. For example, in the absence of perfect price discrimination, the monopolist sets lower-quality (0.25 instead of 0.375), higher prices (0.125 instead of 0.094) and serves a smaller share of the market (50% instead of 75%), which yields only 59% as much total surplus (0.0625 instead of 0.1055).

3.2 Bilateral Monopoly: Independent (Nonintegrated) Monopolists Upstream and Downstream

Consider now the case when each component is produced by a different independent monopolist. The upstream monopolist produces A and the downstream monopolist produces B. They are sold at prices p_A and p_B, respectively, so that the composite good AB is available at price $p = p_A + p_B$. By the same reasoning as earlier, both the upstream and downstream monopolists will choose to set identical qualities. Therefore, the composite good has quality $q = q_A = q_B$. All consumers with marginal willingness to pay for quality larger than $\theta_1 = p/q$ purchase the good. Profits for the two firms are:

$$\Pi_A(p, q) = p_A(1 - \theta_1) - q^2/4 \text{ and } \Pi_B(p, q) = p_B(1 - \theta_1) - q^2/4. \quad (15)$$

Solving the First Order Necessary Conditions (FONCs) for profit maximization for the second stage, $\partial\Pi_1/\partial p_A = 0$ and $\partial\theta_2/\partial p_B = 0$, yields equilibrium prices $p_A^*(q) = p_B^*(q) = q/3$ and $p^*(q) = 2q/3$.[19] When compared to the earlier case of a single integrated monopolist, prices here are higher for any level of quality. This is because of the double marginalization effects first observed by Cournot (1838/1927). Essentially, each of the independent monopolists is unable to appropriate the full benefits of a decrease in its own price.[20]

Anticipating these equilibrium prices, each firm chooses in the first stage the quality of its component so as to maximize its profits:

$$\Pi_1(p^*(q), q) = \Pi_2(p^*(q), q) = q/9 - q^2/4. \quad (16)$$

The noncooperative equilibrium choice for each firm is to set $q^* = 2/9$, which implies that $p^* = 4/27$, $\theta_1 = 2/3$, and $\Pi_1(p^*, q^*) = \Pi_2(p^*, q^*) = 1/81$. Consumer surplus in this case is also $1/81$ ($\cong 0.0123$).

Compared to the case of integrated monopoly, because of double marginalization, in bilateral monopoly marginal increases in quality have a bigger impact on price. Being able to sell the same quality at a higher price than under integrated monopoly, the bilateral monopolists choose lower quality levels, which are less costly. Despite that, because of double marginalization, prices are higher than in integrated monopoly, a lower portion of the market is served, and firms realize lower profits. Consumers also receive lower surplus in comparison to vertically integrated monopoly. The effects of the lack of vertical integration on price are known. The interesting result here is that *a lack of vertical integration (unbundling) leads to a reduction in quality.* Note that this is not because of a lack of coordination between the bilateral monopolists in the choice of quality, because they both choose the same quality level.

These comparisons generalize. Economides (1994) shows that quality is lower in bilateral monopoly than in integrated monopoly *for any distribution of preferences and any cost function.* The same paper shows that, despite the provision of *lower quality,* bilateral monopolists charge a *higher price* for the composite good, as long as the fixed costs of quality are a power function of the quality level.

3.3 Integrated Firm Facing Competition in One Component

We now consider a duopoly in which Firm 1 provides end-to-end service, whereas Firm 2 produces only the downstream component. Here, there are two possibilities we must consider: (a) Firm 1 produces two complementary components (A_1

and B_1) of equal quality, whereas Firm 2 produces product B_2 of lower quality; or (b) Firm 1 produces a high-quality A_1 and a low-quality B_1, whereas Firm 2 produces a high-quality B_2.[21] For each of these cases, there are two additional subcases, depending on whether we allow Firm 1 to price discriminate with respect to the price of A_1 and depending on whether it is bundled with B_1 or B_2.

3.3.1 Integrated Firm Produces Components of the Same Quality

We first consider the case in which both of the components (A_1 and B_1) that are produced by the integrated firm are of the same quality q_1 and are sold at prices w_1 and v_1, respectively. Firm 2 produces good B_2 of quality q_2 and sells it at price v_2. Because demand for Firm 2's good depends on the $\min(q_1, q_2)$, Firm 2 will always choose $q_2 \leq q_1$.

Now, if Firm 1 may price discriminate, it can always set the price of w_1 to hybrid system purchasers sufficiently high to foreclose Firm 2 from the market. Because monopoly profits are higher with a single product (as we have shown earlier), Firm 1 will choose to foreclose Firm 2 and act as a monopolist (as in section 3.1) if it is allowed to price discriminate.

If Firm 1 is prevented from price discriminating, then we need to examine two cases. First, if $q_2 = q_1$, the two systems are perfect substitutes and all of the sales will go to the system with the lower downstream component price, $\min(v_1, v_2)$. At the pricing stage, this will lead to marginal cost pricing for the downstream components, or $v_1 = v_2 = 0$, which will imply negative profits for Firm 2 for any $q_2 > 0$. Therefore, if Firm 2 is to compete, it must choose $q_2 < q_1$ and $v_2 < v_1$.

With this ordering of qualities, demand for A_1B_1 is $1 - \theta_1$ and demand for A_1B_2 is $\theta_1 - \theta_2$, where $\theta_1 = (v_1 - v_2)/(q_1 - q_2)$ and $\theta_2 = (w_1 + v_2)/q_2$. The firms' profits are:

$$\Pi_1 = w_1(1 - \theta_2) + v_1(1 - \theta_1) - q_1^2/2 \text{ and } \Pi_2 = v_2(\theta_1 - \theta_2) - q_2^2/4 \quad (17)$$

The price equilibrium in the last stage is[22] $w_1^* = q_2/2$, $v_1^* = (q_1 - q_2)/2$, and $v_2^* = 0$. Therefore, even if Firm 1 is prohibited from price discriminating, Firm 1 will price at the second stage so as to foreclose Firm 2 from the market. Anticipating this, the second firm would set its quality $q_2^* = 0$ and the first firm would behave as a single product monopolist (section 3.1).

3.3.2 Integrated Firm Produces Components of Different Qualities

We now consider the case in which the components of Firm 1 (A_1 and B_1) are of different qualities q_{A1} and q_{B1}, respectively. Obviously, the bottleneck upstream component must have the higher quality (i.e., $q_{A1} > q_{B1}$). For it to be rational for Firm 1 to produce an upstream component of higher quality, the quality of Firm 2's component, B_2, must be higher than the quality of B_1 (i.e., $q_{B2} > q_{B1}$). Because there is no gain to either firm from producing higher than necessary quality, $q_{A1} = q_{B2} = q_1 > q_{B1} = q_2$.

In the absence of price discrimination, let the prices for A_1, B_1, and B_2 be w_1, v_1, and v_2, respectively. Thus, there is a high-quality hybrid system A_1B_2 with quality q_1 sold at price $w_1 + v_2$ and a low-quality bundled system A_1B_1 with quality q_2 sold at price $w_1 + v_1$. The demands for A_1B_2 and A_1B_1 are $1 - \theta_1$ and $\theta_1 - \theta_2$, respectively, where $\theta_1 = (v_2 - v_1)/(q_1 - q_2)$ and $\theta_2 = (w_1 + v_1)/q_2$. The profit functions are:

$$\Pi_1 = w_1(1 - \theta_2) + v_1(\theta_1 - \theta_2) - q_1^2/4 - q_2^2/4$$
$$\Pi_2 = v_2(1 - \theta_1) - q_1^2/4. \tag{18}$$

Solving for the equilibrium second stage prices yields[23]:

$$w_1^* = (2q_1 + q_2)/6, \quad v_1^* = (q_2 - q_1)/3, \quad v^{2*} = (q_1 - q_2)/3.$$

This solution implies $v_1^* < 0$, which cannot be an equilibrium. Constraining $v_1^* = 0$ implies $q_2^* = 0$ as the quality choice. Therefore, this case reduces to the case of two nonintegrated monopolists (case 3.2).

If we allowed price discrimination, then the previous solution implies equilibrium system prices of $w_1^* + v_1^* = q_2/2$ for A_1B_1 and $w_1^* + v_2^* = (4q_1 - q_2)/6$ for A_1B_2. This implies that $\theta_1 = 2/3$ and $\theta_2 = 1/2$ regardless of the actual qualities chosen. At these prices, firm profits are:

$$\Pi_1(q_1, q_2) = q_1/9 - q_1^2/4 + 5q_2/36 - q_2^2/4 \text{ and}$$
$$\Pi_2(q_1, q_2) = (q_1 - q_2)/9 - q_1^2/4. \tag{19}$$

Solving the first-order necessary conditions[24] yields the following equilibrium quality choices—$q_1^* = 2/9$ and $q_2^* = 5/18$—which is impossible under the assumptions of the case. As q_2 approaches q_1, v_2^* goes to zero and Firm 2's profits become negative. Therefore, regardless of whether price discrimination is allowed or not, there does not exist a quality equilibrium in which the integrated producer has unequal qualities and is competing against a nonintegrated competitor. Furthermore, this implies that the case of two nonintegrated monopolists discussed in section 3.2 is stable against sequential entry in the form of forward (or backward) integration by either of the incumbents via a quality-differentiated component.

3.4 Each of Three Components Produced by a Different Firm

Now consider the case in which there are three firms, each producing a single component. Let Firm 1 be the sole producer of the upstream component A_1, and let Firms 2 and 3 produce competing versions of the downstream component

B_2 and B_3. Let w_1 designate the price of A_1 in the absence of price discrimination; and, when price discrimination is allowed, let w_{12} and w_{13} designate the price of A_1 when bundled with B_2 or B_3, respectively. Let v_2 and v_3 designate the prices of B_2 and B_3.

Once again, a moment's reflection makes it clear that at any equilibrium we must have $q_{A1} = \max(q_{B2}, q_{B3})$, because the first firm will increase its costs but not its revenues if it sets its quality higher than the higher of the two B-component qualities. Without loss of generality, we may assume that $q_1 = q_{A1} = q_{B2} \geq q_{B3}$. This leaves us with two cases to consider: $q_{B2} = q_{B3}$ or $q_{B2} > q_{B3}$. In the former case, the downstream firms' components are perfect substitutes and hence the equilibrium prices will be zero, implying that $q_{B2} = q_{B3} = 0$. Firm 1's best response is to set $q_{A1} = 0$ and no one will purchase anything.

Assuming $q_1 > q_{B3}$, demand for A_1B_2 is $1 - \theta_1$ and demand for A_1B_3 is $\theta_1 - \theta_2$, where $\theta_1 = (v_2 - v_3)/(q_1 - q_{B3})$ and $\theta_2 = (w_1 + v_3)/q_{B3}$ if price discrimination is prohibited, and $\theta_1 = (w_{12} - w_{13} + v_2 - v_3)/(q_1 - q_{B3})$ and $\theta_2 = (w_{13} + v_3)/q_{B3}$, if price discrimination is allowed. In the absence of price discrimination, the profits are:

$$\Pi_1 = w_1(1 - \theta_2) - q_1^2/4 = w_1(1 - (w_1 + v_3)/q_{B3}) - q_1^2/4$$
$$\Pi_2 = v_2(1 - \theta_1) - q_1^2/4 = v_2(1 - (v_2 - v_3)/(q_1 - q_{B3})) - q_1^2/4 \qquad (20)$$
$$\Pi_3 = v_3(\theta_1 - \theta_2) - q_{B3}^2/4 = v_3((v_2 - v_3)/(q_1 - q_{B3}) - (w_1 + v_3)/q_{B3}) - q_{B3}^2/4$$

which implies that the second-stage equilibrium prices are $w_1^* = q_{B3}/2$, $v_2^* = (q_1 - q_{B3})/2$ and $v_3^* = 0$.[25] Therefore, the third firm is forced to charge zero price and chooses in the earlier stage to produce a quality of level zero. But, this is equivalent to Firm 3 dropping out of the market, because the outside good is of quality-level zero. Firms 1 and 2 also anticipate the pricing equilibrium and see that Firm 3 is not a threat to them. Thus, at the first stage, Firms 1 and 2 choose the quality levels of bilateral monopoly (section 3.2). The price levels of bilateral monopoly follow.[26] This outcome can be thought of as a foreclosure of the potential entrant.

If price discrimination is allowed, however, Firm 1's profits become:

$$\Pi_1 = w_{12}(1 - \theta_1) + w_{13}(\theta_1 - \theta_2) - q_1^2/4 \qquad (21)$$

and the new pricing equilibrium is given by:

$$w_{12}^* = q_1(3q_1 + q_{B3})/(9q_1 - q_{B3}), \quad w_{13}^* = 4q_1q_{B3}/(9q_1 - q_{B3}),$$
$$v_2^* = 3q_1(q_1 - q_{B3})/(9q_1 - q_{B3}), \quad v_3^* = q_{B3}(q_1 - q_{B3})/(9q_1 - q_{B3}).$$

These prices yield the following first-stage profits:

$$\Pi_1(p^*(q), q) = (9q_1^3 + 7q_1^2q_{B3})/(9q_1 - q_{B3})^2 - q_1^2/4,$$
$$\Pi_2(p^*(q), q) = 9q_1^2(q_1 - q_{B3})/(9q_1 - q_{B3})^2 - q_1^2/4,$$
$$\Pi_3(p^*(q), q) = q_1q_3(q_1 - q_3)/(9q_1 - q_{B3})^2 - q_{B3}^2/4. \qquad (22)$$

Solving the first-stage FONC yields equilibrium qualities of $q_1^* = 2/9 = 0.222$ and $q_{B3}^* = 0.0209$.[27] This implies $\theta_1^* = 0.663$, $\theta_2^* = .551$, $w_{12}^* = 0.0771$, $w_{13}^* = 0.0093$, $v_2^* = 0.06774$, and $v_3^* = 0.0021$. Equilibrium profits are 0.01471, 0.01050, and 0.00013 for Firms 1, 2, and 3, respectively, and consumer surplus is 0.0136. This case suggests that in order to sustain competition in the downstream component, price discrimination must be allowed.

3.5 Duopoly Competition Between Two Vertically Integrated Firms

In all of the previous cases, the upstream component was a bottleneck facility. The present case considers what happens when we have two integrated firms competing. The discussion of this case is divided into two major subsections. In the first subsection we examine what happens when the components are incompatible so interconnection is not possible and there are no hybrid systems. In the second subsection we assume firms can make their products compatible and interconnection is feasible. In this section we show that firms will price so as to foreclose hybrid systems.

3.5.1 Duopoly Competition of Vertically Integrated Firms with Incompatible Components

In this case there are two systems available. Firm 1 sells system A_1B_1 of quality q_1 for p_1, and Firm 2 sells the system A_2B_2 of quality q_2 for p_2.[28] Without loss of generality, let $q_1 \geq q_2$. Clearly, equal qualities, $q_1 = q_2$, will never be chosen because in the price subgame competition will drive prices to marginal cost. Thus, we confine our attention to the case in which q_1 is strictly greater than q_2. Demand for A_1B_1 is $1 - \theta_1$ and demand for A_2B_2 is $\theta_1 - \theta_2$, where $\theta_1 = (p_1 - p_2)/(q_1 - q_2)$ and $\theta_2 = p_2/q_2$. This yields the following profit functions:

$$\Pi_1 = p_1(1 - \theta_1) - q_1^2/2 \text{ and } \Pi_2 = p_2(\theta_1 - \theta_2) - q_2^2/2. \quad (23)$$

The second-stage pricing equilibrium[29] is:

$$p_1^*(q_1, q_2) = 2q_1(q_1 - q_2)/(4q_1 - q_2), \; p_2^*(q_1, q_2) = q_2(q_1 - q_2)/(4q_1 - q_2).$$

Using these prices to compute the first-stage profits and then differentiating profits with respect to quality to get the first-order necessary conditions produces a system of fourth-order polynomials that do not have an analytic solution.[30] However, these may be solved numerically, yielding the equilibrium qualities $q_1^* = 0.253$ and $q_2^* = 0.048$. These qualities imply $p_1^* = 0.1077$, $p_2^* = 0.0103$, $\theta_1^* = 0.475$, and $\theta_2^* = 0.213$. The profits for Firm 1 are 0.0244 and for Firm 2 are 0.0015. Consumer surplus is 0.0431.

3.5.2 Duopoly Competition of Integrated Firms
That Produce Compatible Components

If the two networks discussed earlier are interconnected, then the hybrid systems A_1B_2 and A_2B_1 may be sold also. This case corresponds to the case in which both networks have adopted *compatible* technologies so that it is technically feasible to create the hybrid products, and the two networks *are* interconnected so that the hybrid systems are actually sold. If price discrimination is allowed, either firm can choose to destroy the market for hybrid systems by setting the price for unbundled components suitably high. Therefore, pricing behavior may be used to determine whether components are *effectively compatible* (i.e., hybrid systems face nonzero demand in equilibrium).

With four components, there are theoretically over $4! = 24$ possible quality orders that we should consider; however, a little thought makes it clear that there are really only three cases of interest. First, even though there are four component qualities to consider and four systems to choose from (two bundled systems and two hybrid systems), the firms would choose to set the highest qualities of upstream and downstream components equal so there would never be more than three distinct component qualities in the marketplace.[31] Without loss of generality we may assume that A_1 is the highest quality component. Letting the qualities of A_1, B_1, A_2, and B_2 be q_{A1}, q_{B1}, q_{A2}, and q_{B2}, respectively, we assume $q_{A1} \geq q_{A2}$. Moreover, it is obvious that in any first-stage equilibrium, $q_{A1} = \max(q_{B1}, q_{B2})$. This still leaves us with three possible orderings for the component qualities to examine:

(i) $q_{A1} = q_{B1} \geq q_{A2} \geq q_{B2}$ (which is the mirror image of $q_{A1} = q_{B1} \geq q_{B2} \geq q_{A2}$)
(ii) $q_{A1} = q_{A2} \geq q_{B1} \geq q_{B2}$ (which is the mirror image of $q_{A1} = q_{A2} \geq q_{B2} \geq q_{B1}$)
(iii) $q_{A1} = q_{B2} \geq q_{B1} \geq q_{A2}$ (which is the mirror image of $q_{A1} = q_{B2} \geq q_{A2} \geq q_{B1}$)

Each of these cases will be discussed both when price discrimination is allowed and when it is prohibited. If price discrimination is prohibited, let D_{11}, D_{22}, D_{12}, and D_{21} designate the demands for the systems A_1B_1, A_2B_2, A_1B_2, and A_2B_1, and let w_1, v_1, w_2, and v_2 designate the prices for A_1, B_1, A_2, B_2. We can write each firm's second-stage profit function as follows:

$$\Pi_1 = w_1(D_{11} + D_{12}) + v_1(D_{11} + D_{21}) - q_{A1}^2/4 - q_{B1}^2/4,$$
$$\Pi_2 = w_2(D_{22} + D_{21}) + v_2(D_{22} + D_{12}) - q_{A2}^2/4 - q_{B2}^2/4. \tag{24}$$

If price discrimination (mixed bundling) is permitted and we let p_1 and p_2 designate the prices for A_1B_1 and A_2B_2, then the profit functions become:

$$\Pi_1 = p_1 D_{11} + w_1 D_{12} + v_1 D_{21} - q_{A1}^2/4 - q_{B1}^2/4,$$
$$\Pi_2 = p_2 D_{22} + w_2 D_{21} + v_2 D_{12} - q_{A2}^2/4 - q_{B2}^2/4. \tag{25}$$

The demand for each system will depend both on the pricing equilibria and on the order of qualities in each of the cases. Let θ_3 designate the consumer who is indifferent between purchasing nothing and the minimal quality product available, θ_2 designate the consumer who is indifferent between purchasing the minimal and intermediate quality good, and θ_1 designate the consumer who is indifferent between purchasing the intermediate and the highest quality good. We know that $0 \leq \theta_3 \leq \theta_2 \leq \theta_1 \leq 1$, and that the demand for the highest quality product will be $1 - \theta_1$, the demand for the intermediate quality product will be $\theta_1 - \theta_2$, and the demand for the lowest quality product will be $\theta_2 - \theta_3$. If we find that $\theta_i = \theta_{i+1}$, then demand for the lesser quality of the two systems is zero.

Case 3.5.2(i). $q_{A1} = q_{B1} \geq q_{A2} \geq q_{B2}$. In this case, we have an integrated high-quality Firm 1 that is competing against a lower quality integrated Firm 2. Notice that it cannot be an equilibrium for all of the components to have identical qualities because that would result in zero prices for all components and hence zero qualities. A deviation in which one firm produced higher than zero quality would be privately rational and hence all zero qualities would not be an equilibrium. Therefore, at least one of the two inequalities must be strict if there is to be an equilibrium with the ordering of component qualities assumed earlier.

Clearly, the system A_1B_1 has the highest quality and the hybrid system A_2B_1 has either the same or lower intermediate quality, depending on whether $q_{A1} = q_{B1} = q_{A2}$ or $q_{A1} = q_{B1} > q_{A2}$. If the former were true, then components A_1 and A_2 become perfect substitutes, and the only equilibrium price that could prevail would be zero. This cannot be part of a subgame-perfect equilibrium because each firm would prefer to lower the quality of its component of type A and free ride on its competitor's higher quality component of type A.[32] Therefore, if there is an equilibrium in this case it must be true that the first inequality is strict.

Assuming that both inequalities are strict, the highest quality product is A_1B_1, the intermediate quality good is the hybrid system A_2B_1, and the lowest quality products are A_1B_2 and A_2B_2, which have identical quality, q_{B2}. In the absence of price discrimination, it is obvious that in any quality price subgame-perfect equilibrium, the price of a higher quality component must be (weakly) higher than the price of a lower quality component.[33] Therefore, the price of A_1 must be greater than the price of A_2 and the price of B_1 must be greater than the price of B_2. In these circumstances, the second-stage profit functions become:

$$\Pi_1 = w_1(D_{11} + D_{12}) + v_1(D_{11} + D_{21}) - q_{A1}^2/4 - q_{B1}^2/4$$
$$= w_1(1 - \theta_1) + v_1(1 - \theta_2) - q_{A1}^2/2$$
$$\Pi_2 = w_2(D_{22} + D_{21}) + v_2(D_{22} + D_{12}) - q_{A1}^2/4 - q_{B2}^2/4$$
$$= w_2(\theta_1 - \theta_2) + v_2(\theta_2 - \theta_3) - q_{A2}^2/4 - q_{B2}^2/4, \qquad (26)$$

where:

$$\theta_1 = (w_1 - w_2)/(q_{A1} - q_{A2}), \quad \theta_2 = (v_1 - v_2)/(q_{A2} - q_{B2}), \text{ and } \theta_3 = (w_2 + v_2)/q_{B2}.$$

The first-order necessary conditions associated with the second-stage pricing equilibrium imply prices that result in $\theta_1 = \theta_2$ or zero demand for the intermediate quality good in the case in which $q_{A1} = q_{B1} > q_{A2} > q_{B2}$.[34] If this is the pricing equilibrium, then Firm 2's revenues are unaffected by setting quality for A_2 higher than B_2 so Firm 2 would wish to set $q_{A2} = q_{B2}$. This leaves only one more case to consider, $q_{A1} = q_{B1} > q_{A2} = q_{B2}$.

With this ordering of component qualities, there are only two system qualities available: a high-quality system produced by Firm 1 and three versions of a low-quality system (the two hybrid systems and the bundled system from Firm 2). Let $H = w_1 + v_1$ be the equilibrium price for the high-quality system, and $L = \min(w_1 + v_2, w_2 + v_1, w_2 + v_2)$ be the equilibrium price of the low-quality system. Obviously, $H > L$ or we again end up with all zero component qualities, which is not an equilibrium.

We now show that with each firm having two prices to set and with zero marginal costs (at the second stage), there is no positive component pricing equilibrium. Without loss of generality, assume $w_2 \leq v_2$. If $q_{A2} > 0$, then $v_2 > 0$. In this case, Firm 1 would always prefer to deviate to $v_1^\dagger = v_2 - \varepsilon$ and $w_1 = H - v_1^\dagger$. However, $v_1 > 0$ and $v_2 = 0$ (implying that $q_{A2} = q_{B2} = 0$) is not an equilibrium because Firm 2 would always prefer to deviate to $v_2^\dagger = v_1 - \varepsilon$ and $w_2 = L - v_2^\dagger$. Finally, $v_1 = v_2$ is not an equilibrium because this again implies $q_{A1} = q_{B1} = q_{A2} = q_{B2} = 0$. Therefore, there is **no** pricing equilibrium that can support the earlier ordering of qualities if price discrimination is prohibited.

In the second stage, the marginal cost of selling a unit of any quality is zero, and so neither high-quality nor low-quality firms can commit to not succumbing to the Bertrand temptation to cut prices to capture additional revenue in the low-quality market. Permitting price discrimination would allow the firms to set unbundled prices sufficiently high to prevent competition from hybrid system products, which would convert this case into the case of competition between vertically integrated duopolists producing incompatible components.

There is no other equilibrium possible with price discrimination because any case in which there is an active hybrid system market would have at least two products competing that are perfect substitutes and with three pricing instruments, the equilibrium would be vulnerable to similar deviations to those described earlier.

Case 3.5.2(ii). $q_{A1} = q_{A2} \geq q_{B1} \geq q_{B2}$. With this ordering, A_1 and A_2 have identical qualities and, by arguments made previously, it is clear that Firm 1 would also choose to set the same quality for B_1, which leaves us with only one case to consider: $q_{A1} = q_{B1} = q_{A2} > q_{B2}$. Without price discrimination, we again have the problem of both firms wishing to free ride on the other firm's quality investment in A_1, which means that this cannot be an equilibrium ordering.

Case 3.5.2(iii). $q_{A1} = q_{B2} \geq q_{B1} \geq q_{A2}$. Here the highest quality system is the hybrid system A_1B_2, the intermediate quality system is A_1B_1, and the lowest quality system is either A_2B_2 or A_2B_1. Following arguments similar to those made earlier, it is clear that there is no quality or pricing equilibrium with this ordering that faces nonzero demand.

The discussion in this section makes it clear that there is no equilibrium with positive demand for hybrid system products for any of the possible orderings of component qualities. This seems in contrast with previous results on the choice of compatibility by vertically integrated duopolists. Matutes and Regibeau (1988, 1992) and Economides (1989) have shown that, for symmetric demand systems (when the demands for hybrids are equal to the demand for single-producer composite goods at equal prices), vertically integrated firms prefer compatibility. Economides (1988, 1991) has shown that this result can be reversed if the demand for hybrids is small relative to the demand of the own system. This is because compatibility increases demand but also increases competition. Thus, a firm that faces a small hybrid demand (relative to the demand of the own system) does not get rewarded sufficiently in terms of demand for the increased competition brought by compatibility, and therefore it prefers incompatibility. In the model of this chapter, the demand for hybrids depends on the quality choices of all four components. Because the quality of a composite good is the minimum of the qualities of its component parts, any quality configuration in which the components of each vertically integrated firm have the same quality level, but in which this quality level is different than the common quality of the components of the opponent ($q_{A1} = q_{B1} > q_{A2} = q_{B2}$), will lead both firms to desire incompatibility. In such a case, in the price stage, the hybrid does not give enough demand reward to the low-quality producer, who therefore chooses incompatibility. In a case in which an integrated firm produces components of different qualities (e.g., $q_{A1} = q_{B1} > q_{A2} > q_{B2}$), in the pricing stage firms have an incentive to choose compatibility noncooperatively. However, at the earlier quality stage, the firm that was assumed to produce different qualities has an incentive to equalize its quality levels downward. This move leads the market to the earlier case ($q_{A1} = q_{B1} > q_{A2} = q_{B2}$) in which compatibility is not desirable from each firm's point of view. In summary, given certain quality levels, compatibility is desirable as in the previous literature. However, allowing the firms to choose their quality levels drives them toward the market structures in which compatibility is undesirable to them.[35] Essentially, *the drive of integrated firms to achieve a uniform quality of all their components while differentiating their quality from that of opponents squeezes out the demand for hybrids and drives firms to incompatibility and lack of interconnection.*

4. IMPLICATIONS FOR THE TELECOMMUNICATIONS INDUSTRY

The history of the Public Switched Telecommunications Network (PSTN) provides a useful vehicle for interpreting the results derived in the preceding sections. As the PSTN evolves into a mixed, hybrid system of public and private facilities, firms' strategic incentives will change. The forces that are altering the PSTN are quite complex. They include technological progress, globalization, and deregulation. For example, advances in fiber optics, digital switching, and software control are reducing costs and creating new product opportunities. Increased internationalization has encouraged multinational alliances, whereas deregulation is permitting new types of entry. Our model addresses these forces indirectly by examining how changes in the ownership structure of network components and selected regulatory reforms (e.g., ONA pricing and interconnection requirements) change firms' strategic behavior with respect to pricing and product design, which is interpreted as choosing the quality for network services.

The cases summarized in Figs. 1.3 and 1.4 reflect static equilibria that ignore such important real-world issues as multipart tariffs (i.e., separate fees for access and usage), volume discounting to discriminate between residential and commercial customers (e.g., WATS), and special-access or custom-designed virtual private network services. Most importantly, we ignore the effects of rate-of-return regulation.[36] These omissions are intentional. We believe that telecommunications is moving fast in a completely deregulated environment, and it is this environment that now needs to be studied. Thus, we focus attention on noncooperative equilibria in prices and quality levels and examined changes in ownership structure, isolated from the possibly distorting influences of technological change, regulatory politics, and dynamic investment planning. Therefore, the analogies drawn between the real-world situation and the cases are intended to be suggestive rather than exact. However, we believe that our analysis is closer to the near future state of affairs in telecommunications than an analysis based on regulation would be. With this proviso in mind, let us interpret our results in light of current and prospective changes in the structure of the U.S. telecommunications industry.

The first case we considered (section 3.1) was the vertically integrated monopolist. This is our base case and corresponds loosely to the predivestiture world of AT&T's Bell System, in which a single firm provided end-to-end Plain Old Telephone Service (POTS) of a uniform quality as its principal activity. Compared to the socially efficient solution (section 3.1.3), a monopolist sets higher prices, offers lower quality, and serves a smaller share of the market, yielding approximately half of the socially efficient level of surplus.[37] The avowed intention of rate-of-return regulation was to help correct this inefficiency. Unfortunately, regulation is costly and introduces its own distortions.

The divestiture of AT&T was intended to improve total surplus by encouraging competition in the long distance markets. The initial effects of divestiture are captured by a comparison of the first case with the case of bilateral monopolists (section 3.2). Compared to an integrated monopolist, marginal increases in quality have a bigger impact on price. As a result, firms choose lower quality levels, serve a smaller portion of the market, and realize lower profits despite the higher prices. Consumer surplus is lower, also implying that in this model divestiture reduces total welfare and has a negative impact on service quality.

At the time of divesture, AT&T no longer had a monopoly over long distance service. Indeed, the emergence of alternative interexchange carriers such as MCI in the 1970s helped provide the impetus to force the divestiture of long distance and local exchange services. The case of three firms discussed in section 3.2 captures the impact of combining divestiture with increased competition in one of the service markets. If price discrimination is permitted, competition offers an improvement relative to bilateral monopoly, but still falls short of the original vertically integrated monopoly outcome.[38]

If price discrimination is prohibited, an equilibrium with three firms is not sustainable. Two of the firms act to foreclose the third firm. If price discrimination is allowed, however, the firm that controls the bottleneck facility (i.e., the local exchange carrier in the present discussion) finds it advantageous to price so as to encourage entry of another downstream firm. This result suggests that ONA pricing by the local exchange carrier, in the absence of price regulations, would reduce total efficiency. It highlights the close linkage between price discrimination requirements, compatibility standards, quality choices, and the vertical structure of firms in the industry.

Prior to divestiture, AT&T faced competition from nonintegrated long distance carriers such as MCI. MCI originally competed against AT&T with a lower quality network and had to sue AT&T for access to its local exchange services.[39] MCI won its suit. The analysis of the case discussed in section 3.3.1 shows that competition by a lower quality, nonintegrated carrier is not sustainable regardless of whether interconnection (compatibility) is required or whether price discrimination is prohibited (i.e., AT&T was required to price as a common carrier).

Today, the local exchange carriers are agitating for regulatory reforms that would permit the reintegration of local and long distance services.[40] As long as they maintain an effective monopoly over local access services, it may be necessary to regulate access pricing if nonintegrated competitors are to survive.

The second case of nonintegrated competition (section 3.3.2) involves an integrated firm with both a high-quality bottleneck and a low-quality subnetwork competing against a nonintegrated high-quality firm. Here, competition is sustainable only if price discrimination is prohibited, but the outcome is worse than under the monopoly case (i.e., the solution defaults to the bilateral monopoly result). Such a situation might arise if a long distance company attempted to enter local access competition via an alliance with a cable company, an alternative access provider such as Teleport or MFS, or via cellular access (e.g., McCaw and AT&T). At least, initially, we might expect the extent of facilities and the coordination of interconnection to result in the Bell Operating Company being perceived as offering higher quality access facilities.

The results considered here assume the existence of a bottleneck facility. What happens if there are two integrated firms competing as quality-differentiated duopolists? For example, imagine the situation in which AT&T integrated forward into local exchange services via cellular or cable TV access while the local exchange carrier simultaneously integrated backward into long distance services. The alliances between Bell-Atlantic and TCI, Time Warner and US West, MCI and Teleport, AT&T and McCaw Cellular, and so on, suggest that this may be the mode of competition in the future. Fortuitously, the results presented in section 3.2.1 indicate that this will produce an outcome that outperforms the integrated monopolist. It is noteworthy that this is the *only* industry structure that outperforms the original predivestiture model, and that even this solution falls far short of the socially optimal level of total surplus (i.e., total surplus with integrated duopolists is 0.069 versus 0.1055). Thus, even with duopoly competition in every market, continued regulation may be justified. Moreover, our results assume that the production costs are such that duopoly competition is sustainable (i.e., local access is *not* a natural monopoly).

Our model indicates that there are no gains from offering hybrid systems. The integrated firms will price so as to foreclose these markets. This may be unfortunate because other reasons to interconnect might exist. For example, active hybrid system markets may be useful in helping to assure network reliability (i.e., hybrid systems offer alternative routing options in the event one of the subnetworks fails). However, because real networks are unlikely to perfectly overlap, the opportunity to reach additional customers will help encourage interconnection.

5. CONCLUDING REMARKS

Although the preceding discussion attempted to interpret our analytic results in terms of policy questions that are of interest to telecommunications regulators, we should not forget that this is an abstract model. Its real contribution may be in expanding our understanding of the theoretical tools that underlie economic-based policy analysis. We were somewhat surprised to find that the standard models were inadequate for addressing the questions posed by the evolution toward a more highly decentralized information infrastructure.

Much work needs to be done if we are to understand the effects of changing industry structure on incentives to invest in quality. For example, two important theoretical extensions to the present work include (a) accounting for the effect of positive externalities on quality investments, and (b) incorporating imperfect, asymmetric information. In addition to these theoretical extensions, we need additional empirical work that addresses how the quality of our infrastructure has changed since divestiture.

ENDNOTES

1. For a comprehensive discussion of the economics of networks, see Economides and White (1994).
2. Moreover, the quality of the bottleneck facility sets the maximum system quality that will be available to consumers. This last point follows directly from our modeling assumption that the quality of a system is never higher than the quality of the weakest link in the system.

3. Although the simplified approach toward costs in the present analysis includes determining whether such competition is feasible (e.g., whether local access services are a natural monopoly), it helps alleviate our concern regarding the strategic impact of competition on infrastructure quality.

4. The present analysis is inappropriate when such a ranking is impossible. This may occur if consumers measure quality along multiple dimensions that are not mappable into a unitary index (e.g., they view a comparison of reliability and customer service as akin to comparing "apples and oranges").

5. See, for example, Besen and Johnson (1986), David and Greenstein (1990), Economides and White (1994), Farrell and Saloner (1985, 1986), and Katz and Shapiro (1986).

6. Gabszewicz and Thisse (1979, 1982), Mussa and Rosen (1978), Shaked and Sutton (1982), among others.

7. As is typical, we assume perfect information and symmetric production technologies. Consumers know the prices and qualities of all available products when they make their purchase decisions and firms agree on the nature of consumer demand and the structure of the strategic game they are playing. Symmetric production technologies assure that the firms' costs are similar, which focuses attention on their strategic behavior with respect to product design and pricing.

8. p_L must be less than p_H or the demand for the low-quality product will be zero.

9. They may also charge different component prices based on the identity of the firm supplying the complementary component if there is more than one.

10. This latter assumption could be relaxed if we reinterpret prices as increments above a constant positive marginal cost.

11. At equilibrium, no firm can increase its profits by altering its behavior unilaterally, given that the other firms are playing their Nash strategies. Subgame perfection ensures that the equilibrium is appropriately decentralizable in any subgame.

12. The ordering of component qualities will define an ordering for product qualities. For example (see section 3.5.2), with two firms each producing an upstream and downstream product that are compatible, there are four system products and over $4! = 24$ possible orderings for the component qualities. Once we know the ordering of product qualities we can infer consumer demand as a function of prices and qualities and can write down the firms' profit functions. The First Order Necessary Conditions (FONCs) obtained when profits are differentiated with respect to prices identify the pure strategy equilibrium for the second stage.

13. To be active a firm must offer a component with positive quality. Because this incurs a fixed entry cost that is sunk in the second stage, there is no free entry in the pricing game. This precludes "hit-and-run" entry which, in any case, seems unlikely in telecommunications.

14. We ignore cases that are only the mirror image of one already considered and dispense with trivial cases in a sentence or two.

15. Integrating consumer surplus of individual types $U_\theta(q^*, p^*) = \theta q^* - p^*$ over $[\theta_1, 1]$ we get $\frac{1}{2}(1 - \theta_1^2)q^* - (1 - \theta_1)p^* = 1/32 = 0.03125$.

16. Substituting for θ_1 and θ_2 and differentiating (11) with respect to p_1 and p_2, respectively, yields the following two FONCs: $((q_1 - q_2) - 2(p_1 - p_2))/(q_1 - q_2) = 0$ and $2(q_2 p_1 - q_1 p_2)/(q_2(q_1 - q_2)) = 0$. Solving these for p_1 and p_2 yields the indicated solution.

17. Because costs do not increase but total surplus does increase when demand increases, it is efficient to price discriminate so that everyone consumes the highest quality product. Because no one would consumer lower quality products, these should not be produced.

18. From section 3.1.2, we know that the monopolist does not find this profitable in our model. We did not compute the socially optimal solution under uniform pricing with multiple products because the incremental gains over the single-product uniform case would be small (i.e., the socially optimal single-product/uniform-pricing solution already produces a surplus equal to 84% of the level achieved with perfect price discrimination).

19. This symmetric pricing equilibrium is the unique Bertrand pricing equilibrium.

20. See Economides and Salop (1992) for a discussion of the effects of double marginalization of vertical mergers under compatibility.

21. The other possibilities are not viable. If all three qualities are equal then the composite goods have the same quality, which leads to marginal cost pricing and is not an equilibrium. If the upstream component A_1 is low quality, then again the composite goods will have the same (low) quality.

22. First-order conditions are $0 = \partial\Pi_1/\partial w_1 = 1 - (2w_1 + v_2)/q_2$, $0 = \partial\Pi_1/\partial v_1 = 1 - (2v_1 - v_2)/(q_1 - q_2)$, $0 = \partial\Pi_2/\partial v_2 = (v_1 - 2v_2)/(q_1 - q_2) - (w_1 + 2v_2)/q_2$.

23. The FONCs are $\partial\Pi_1/\partial w_1 = (q_1 - 2(w_1 + v_1))/q_2 = 0$, $\partial\Pi_1/\partial v_1 = 2(w_1 + v_1)/q_1 + (2v_1 - v_2)/(q_1 - q_2) = 0$, and $\partial\Pi_2/\partial v_2 = 1 + (v_1 - v_2)/(q_1 - q_2) = 0$.

24. The FONCs are $\partial\Pi_1/\partial q_1 = \partial\Pi_2/\partial q_2 = 1/9 - q_1/2 = 0$ and $\partial\Pi_1/\partial q_2 = 5/36 - q_2/2 = 0$.

25. The FONCs are $\partial\Pi_1/\partial q_1 = (q_{B3} - v_3 - 2w_1)/q_{B3} = 0$, $\partial\Pi_2/\partial q_1 = (q_1 - q_{B3} - 2v_2 + v_3)/(q_1 - q_{B3}) = 0$, and $\partial\Pi_3/\partial q_{B3} = (v_2 q_{B3} - 2v_3 q_1 - w_1(q_1 - q_{B3}))/(q_{B3}(q_1 - q_{B3})) = 0$.

26. If Firms 1 and 2 behave as bilateral monopolists and set their qualities at (2/9) and Firm 3 chooses to enter with a quality greater than zero, Firms 1 and 2 will price to foreclose Firm 3 from the market; thus the bilateral monopoly solution is a Nash equilibrium in the three-firm game without price discrimination.

27. The FONCs are $\partial\Pi_1/\partial q_1 = q_1(162q_1^2 - 729q_1^3 - 54q_1 q_{B3} + 243q_1^2 q_{B3} - 28q_{B3}^2 - 27q_1 q_{B3}^2 + q_{B3}^3)/(2(9q_1 - q_{B3})^3) = 0$, $\partial\Pi_2/\partial q_1 = q_1(162q_1^2 - 729q_1^3 - 54q_1 q_{B3} + 243q_1^2 q_{B3} + 36q_{B3}^2 - 27q_1 q_{B3}^2 + q_{B3}^3)/(2(9q_1 - q_{B3})^3) = 0$, and $\partial\Pi_3/\partial q_{B3} = (18q_1^3 - 34q_1^2 q_{B3} - 729q_1^3 q_{B3} + 243q_1^2 q_{B3}^2 - 27q_1 q_{B3}^3 + q_{B3}^4)/(2(9q_1 - q_{B3})^3) = 0$. This system of equations was solved numerically to yield the indicated result.

28. Once again, it is obvious that $q_1 = q_{A1} = q_{B1}$ and $q_2 = q_{A2} = q_{B2}$.

29. The FONCs are $\partial\Pi_1/\partial p_1 = 0 = 1 - (2p_1 - p_2)/(q_1 - q_2)$ and $\partial\Pi_2/\partial p_2 = 0 = (q_2 p_1 - 2q_1 p_2)/(q_2(q_1 - q_2))$.

30. The FONCs are as follows: $\partial\Pi_1/\partial q_1 = (q_1(16q_1^2 - 64q_1^3 - 12q_1 q_2 + 48q_1^2 q_2 + 8q_2^2 - 12q_1 q_2^2 + q_2^3))/(4q_1 - q_2)^3 = 0$, and $\partial\Pi_2/\partial q_2 = (4q_2^3 - 7q_1^2 q_2 - 64q_1^3 q_2 + 48q_1^2 q_2^2 - 12q_1 q_2^3 + q_2^4)/(4q_1 - q_2)^3 = 0$.

31. It never makes sense to set the quality of a component higher than the quality of the highest quality component with which it may be bundled. Therefore, the highest quality A component and the highest quality B component will have the same quality in any quality equilibrium. If the remaining three components all have different and lower qualities, combining these components yields at most an additional intermediate and low-quality system.

32. It cannot be an equilibrium for $q_{A1} = q_{A2} = 0$ because this implies $q_{B2} = 0$, which is not an equilibrium as discussed earlier. Therefore, the deviation to lower quality by one firm is always possible. This deviation is privately attractive because the firm's revenues are unaffected while costs decline. Notice that this is true regardless of the pricing equilibrium as long as price discrimination is prohibited.

33. Were this not the case, then the lower quality component would face zero demand. For example, if $v_1 < v_2$, then sales of B_2 would be zero. This could not be a pricing equilibrium because Firm 2 could deviate to $v_2 = v_1 - \varepsilon$ and capture all of the low-quality B market, which would weakly improve Firm 2's revenues without affecting its costs. In any profit maximizing equilibrium of the game, the lower quality (lower cost) firm can always pursue such a pricing strategy.

34. The FONCs are as follows: $\partial\Pi_1/\partial w_1 = 1 - (2w_1 - w_2)/(q_{A1} - q_{A2}) = 0$, $\partial\Pi_1/\partial v_1 = 1 - (2v_1 - v_2)/(q_{A2} - q_{B2}) = 0$, $\partial\Pi_2/\partial w_2 = (2w_2 + v_2))/q_{B2} + (w_1 - 2w_2))/(q_{A1} - q_{A2}) = 0$, $\partial\Pi_2/\partial v_2 = 0 = 1/q_{B2} + [q_{B2}(2w_2 + v_1) - 2q_{A2}(w_2 + v_2)]/q_{B2}(q_{A2} - q_{B2}))$. The solution of these FONCs yields the following prices: $w_1 = [(2q_{A1})(q_{A1} - q_{A2})]/\lambda$, $w_2 = [(q_{B2}(q_{A1} - q_{A2})]/\lambda$, $v_1 = [2q_{A1}(q_{A2} - q_{B2})]/\lambda$, and $v_2 = [2q_{B2}(q_{A2} - q_{B2})]/\lambda$, where $\lambda = 4q_{A1} - q_{B2}$.

35. This opens the possibility that allowing firms to endogenously choose their quality or variety levels may reverse the compatibility results of Matutes and Regibeau (1988, 1992) and Economides (1989) in general. This could happen if firms choose variety or quality levels that

lead to configurations in which the demand for hybrids is small so that compatibility is undesirable, as in Economides (1989, 1991).

36. Rate-of-return regulation may encourage excessive investments in quality if quality improvements are capital intensive (Averch & Johnson, 1962), which seems plausible (e.g., redundant
electronics to improve reliability, excess channel capacity to reduce blockage, or additional
software features).

37. From Fig. 1.4, total surplus under a vertically integrated monopolist is 0.063, which is 59% of
the socially efficient level of 0.106 (if we are constrained to uniform pricing) and 50% of the
overall efficient level of 0.125 (if perfect discrimination is feasible).

38. From Fig. 1.3, Firm 1's profits increase from 0.01235, under the bilateral monopoly solution, to
0.0244 (section 3.4 with discrimination).

39. MCI's network was initially smaller and its access connections were of lower quality than those
of AT&T.

40. The separation of local and long distance services is required by the Modified Final Judgment,
which is the consent decree between AT&T and the Justice Department governing the divestiture terms.

REFERENCES

Averch, H. and Leland Johnson, (1962), "Behavior of the Firm Under Regulatory Constraint,"
American Economic Review, vol. 52, 1053–1069.

Baumol, William, John Panzar and Robert Willig (1982), *Contestable Markets and the Theory of
Industry Structure*, New York: Harcourt, Brace and Jovanovich.

Besen, Stanley M. and Leland Johnson, (1986), *Compatibility Standards, Competition, and Innovation
in the Broadcasting Industry*, Rand Corporation, R-3453-NSF.

Cournot, Augustin, (1927), *Researches into the Mathematical Principles of the Theory of Wealth*, (N.T.
Bacon Trans.), New York: Macmillan 1927. (Original work published 1838).

David, Paul and Shane Greenstein, (1990), "The Economics of Compatibility Standards: An
Introduction to Recent Research," *Economics of Innovation and New Technology*, vol. 1, pp. 3–41.

Economides, Nicholas, (1988), "Variable Compatibility without Network Externalities," Discussion
Paper No. 145, Studies in Industry Economics, Stanford University.

Economides, Nicholas, (1989), "Desirability of Compatibility in the Absence of Network
Externalities," *American Economic Review*, vol. 78, no. 1, pp. 108–121.

Economides, Nicholas, (1991), "Compatibility and the Creation of Shared Networks," in *Electronic
Services Networks: A Business and Public Policy Challenge*, edited by Margaret Guerin-Calvert and
Steven Wildman, New York: Praeger Publishing Inc., 1991.

Economides, Nicholas, and Steven C. Salop, (1992), "Competition and Integration among
Complements, and Network Market Structure," *Journal of Industrial Economics*, vol. XL, no. 1,
pp. 105–123.

Economides, Nicholas and Lawrence J. White, (1994), "Networks and Compatibility: Implications
for Antitrust," *European Economic Review*, vol. 38, March 1994, pp. 651–662.

Farrell, Joseph and Saloner, Garth (1985), "Standardization, Compatibility, and Innovation," *Rand
Journal of Economics*, vol. 16, pp. 70–83.

Farrell, Joseph and Saloner, Garth, (1986), "Installed Base and Compatibility: Innovation, Product
Preannouncement, and Predation," *American Economic Review*, vol. 76, pp. 940–955.

Hotelling, Harold, (1929), "Stability in Competition," *Economic Journal*, vol. 39, pp. 41–57.

Jaskold-Gabszewicz, Jean and Jacques-Francois Thisse, (1979), "Price Competition, Quality, and
Income Disparities," *Journal of Economic Theory*, vol. 20, pp. 340–359.

Jaskold-Gabszewicz, Jean and Jacques-Francois Thisse, (1980), "Entry (and Exit) in a Differentiated
Industry," *Journal of Economic Theory*, vol. 22, pp. 327–338. Reprinted in Andreu Mas-Colell,

(ed.), *Noncooperative Approaches to the Theory of Perfect Competition* (1982), New York: Academic Press.

Matutes, Carmen, and Pierre Regibeau, (1988), "Mix and Match: Product Compatibility Without Network Externalities," *Rand Journal of Economics*, vol. 19 (2), pp. 219–234.

Matutes, Carmen, and Pierre Regibeau, (1992), "Compatibility and Bundling of Complementary Goods in a Duopoly," *Journal of Industrial Economics*, vol. XL, no.1, pp. 37–54.

Mitchell, Bridger and Ingo Vogelsang, (1991), *Telecommunications Pricing: Theory and Practice*, New York: Cambridge University Press.

Mussa, Michael and Sherwin Rosen, (1978), "Monopoly and Product Quality," *Journal of Economic Theory*, vol. 18, pp. 301–317.

Shaked, Avner and John Sutton, (1982), "Relaxing Price Competition Through Product Differentiation," *Review of Economic Studies*, vol. 49, pp. 3–14.

Shaked, Avner and John Sutton, (1983), "Natural Oligopolies," *Econometrica*, vol. 51, pp. 1469–1484.

Chapter 2

The Political Economy of Entry Into Local Exchange Markets

Bhaskar Chakravorti
Yossef Spiegel
Tel Aviv University

1. INTRODUCTION

Regulated monopolies are constantly under the threat of entry; often regulatory protection constitutes the only barrier to potential entrants. One explanation for such protection derives from Posner's (1971) view of regulation as a taxation scheme. According to this view, regulators set prices as if the overall market were divisible into two (or more) segments: In one segment prices are set artificially high in order to generate profits, which are then used to subsidize the other segment in which prices are set artificially low. The regulator's incentive to thus tax one segment and subsidize the other could be driven by a variety of motives: a sense of equity, a desire to ensure universal service, or by the fact that the subsidized segment constitutes a stronger political base.

In such an environment, an entrant would appear to be an anathema from the regulator's standpoint; the (unregulated) entrant would be attracted only to the market in which the regulated price is artificially high so subsequent to entry the regulated firm's profits from this market would decline, thereby eroding the "tax base" and reducing the social cross-subsidy.

The basic elements of the scenario just presented conform well with the local exchange market of today. Rates charged to one set of consumers are abnormally high, whereas another set pays a price below the cost of the service it receives.[1] A U.S. Telephone Association study argued that the subsidies that are built into the current pricing system of phone services totaled more than $20 billion a year.[2]

43

A typical example of such segmentation is between *commercial* and *residential* consumers. Palmer (1992) uses data from New England Telephone to test for cross-subsidization between commercial and residential consumers. She finds that the former cross-subsidizes the latter at 54% of all central offices in her sample (with nearly 65% of all suburban central offices exhibiting this phenomenon). The average minimum monthly subsidy contribution per commercial line for central offices in suburban areas is approximately $6.41, whereas the average minimum monthly subsidy benefit per residential line is $2.22. In fact, FCC data show that the national average rate for residential unlimited local telephone service for 1992 was $13.08, compared with $32.38 for single-line unlimited service for businesses. Similarly, the national average basic residential connection charge in 1992 was $41.51, compared with $72.61 for businesses.[3]

Other forms of segmentation for the purposes of cross-subsidization are also evident. Daniel, Shin, and Ward (1993) report that in a comment filed with the Federal Trade Commission (FTC) in March 1993, Federal Communications Commission (FCC) staff members concluded that the price of local service was less than half its long-run marginal cost, whereas the price of toll service was at least twice its long-run marginal cost.

A classic case study is that of Pacific Bell, which offers one of the highest subsidies for local rates in the United States. The basic service fee for residential consumers is $8.35 per month, whereas the marginal cost of providing basic local service is $22 per month, according to Pacific Bell estimates. The subsidies are funded by artificially high intraLATA toll charges, access charges, and rates for the Yellow Pages. An average intraLATA toll call in California costs 20 cents, whereas the allocated direct cost of the call is merely 5–6 cents. The California Public Utilities Commission is considering opening up the local exchange market to entrants. Pacific Bell has proposed that toll rates be brought down so that an average intraLATA toll call would cost 8–9 cents; it estimates that it is likely to lose over $100 million of toll business to the emerging competition. On the other hand, local residential rates are proposed to be increased to $13.35 per month. Although it is unlikely that all of the proposals put forward by Pacific Bell regarding the redesigning of the rate structure will be accepted, it is quite clear that there is a fundamental tension from the regulator's standpoint between promoting competition and maintaining local cross-subsidies.

In light of the Posnerian view of regulation as a taxation scheme, it is indeed somewhat surprising to observe the rate at which regulatory barriers to entry in local exchange markets are being dismantled. Entry has taken a variety of forms; the competitors to Local Exchange Companies (LECs) include interexchange carriers, competitive access providers (CAPs), cable TV companies, and wireless and satellite-based communications. The objective of this chapter is to investigate the reasons that may explain this apparent anomaly and their implications for social welfare and for the variety of services available to consumers.

We identify two factors that are crucial to the explanation. The first is the demand for enhanced "quality" of services. For instance, it is no longer the case that all telephone connections are the same. New technologies have made a menu of additional services possible. In addition, given the complexity of the modern telephone network, and the breadth and scope of its uses, the deployment and maintenance of both the relevant hardware and associated software provide a wide range in the quality of service offerings.

The second factor is that of differentiation between the services provided by the regulated firm and by an entrant. A key aspect of the competition in local exchange is that the firms do not compete exclusively on price. Competitive access providers (CAPs) promote their services by offering to connect commercial customers with high-speed lines in a short period of time, greater degree of reliability, and route diversity. These features differentiate the CAPs' services from those of the incumbent firm; the latter have other characteristics to offer such as an established reputation, and brand name recognition.

Our basic model rests on the following arguments. The regulator is modeled as an entity whose objective function depends on the welfare of two classes of consumers, which, for the sake of convenience, we label *residential* and *commercial*. The regulator places a relatively higher weight on the welfare of the residential consumers. The regulated firm is allowed to charge prices in the commercial market that could potentially exceed the marginal cost of service, and a proportion of the profits are transferred to the residential consumers in the form of a price subsidy. Larger profits yield larger transfers and increase the utility of residential consumers. On the other hand, if the regulated commercial price is set high, then the utility of residential consumers is diminished.

A key distinction between the commercial and the residential consumers in our model is that the former derive utility from the investments that the firm has made in quality-enhancing technologies. Examples of such investments include the deployment of digital-switching equipment, fiber-optic cables, and enhanced software capabilities for increased reliability, repair records, and route diversity. Such enhancements permit, for example, the provision of high-speed private lines that can carry up to 672 voice channels, video conferencing, lines that can carry high-volume data traffic, and non-dial-up features of private lines that are particularly attractive to financial trading firms. We argue that given the difficulties of defining "enhanced" quality in a subjective and unambiguous manner, and given the difficulties of monitoring its provision, regulators may be unable to mandate the level of quality-enhancing investments directly. Thus, regulators would only be able to use prices as a way to induce optimal investment. But, because prices are also used to generate social cross-subsidies, they will not lead to an optimal investment. In equilibrium, the firm will invest too little from the regulator's perspective. This underinvestment problem provides regulators with an incentive to invite entrants, in the hope that competition would drive up such investments.

Entry, however, comes at a cost: The regulated firm's profits in the commercial market are diminished, which in turn erodes the base from which transfers are made to the residential market. A priori, the benefit of entry is clearly the positive effect on investment in quality. The regulator's incentive to permit entry and the resulting change in the market structure is therefore derived from an evaluation of this trade-off. This trade-off is apparent from the statement of Reed Hundt, Chairman of the FCC, before the U.S. House of Representatives Subcommittee on Telecommunication and Finance: "Enthusiasm for promoting new competitive markets and encouraging new technologies and services must not distract our attention from the critical task of ensuring that all Americans have access to basic telephone service."[4]

However, there is an additional element that profoundly affects the regulator's decisions. As mentioned previously, typically, the entrant's output is likely to be differentiated from the output of the regulated incumbent. Our analysis characterizes precisely how such differentiation affects the decision to permit entry into a regulated market.

In the following section we provide a brief review of the local telephone industry that has motivated our study. The review is selective in the sense that we focus on the aspects of the industry that pertain to the specific question that we seek to address.

2. INSTITUTIONAL BACKGROUND

2.1 LECs and CAPs

The local telephone industry is dominated by the seven Regional Bell Operating Companies (RBOCs) and GTE. The RBOCs collectively account for 76.7% of the total access lines in the United States in 1992 and about 80% of total revenues from local exchange services.[5] Each of these firms is virtually a monopoly in its own operating region. Potential competitors for these local exchange carriers (LECs) are interexchange carriers (such as AT&T, MCI, and Sprint), cable TV companies, competitive access providers (CAPs; such as Metropolitan Fiber Systems, Teleport), cellular telephone companies, and the next generation in wireless communications, called PCS. Often the lines between these different firms are blurred; a classic example being Teleport Communications Group, which is a CAP that originated as an agreement between Merrill Lynch and AT&T in the mid-1980s and is now jointly owned by several cable TV companies including TCI, Cox Enterprises, and Time-Warner. The threats posed by the various entrants are quite different from each other. They can roughly be put into three categories: network access services, intraLATA (local access and transport area) toll services, and unregulated services such as wireless communications and broadband and cable operations.

Rather than examining the potential rationale and implications for permitting all varieties of entrants, our focus is on the first of the three categories listed earlier—network access services. In 1992, 23.4% of the revenue of an average RBOC was from providing interstate access, and 6.7% of the revenue was from providing intrastate access.[6] All interstate traffic and access rates are regulated by the FCC, whereas intrastate traffic and access rates are regulated by the states' public utility commissions (PUCs). IntraLATA traffic is predominantly regulated by state PUCs. Approximately $13.75 billion out of the $15.75 billion earned by the average RBOC from providing interstate network access in 1992 was from end-user revenues and from switched-access revenues. End-user revenues are FCC subscriber line charges that represent a federally tariffed flat fee independent of usage. In 1992, the subscriber line charge for residential and single-line businesses was $3.49 per month, whereas the charge applicable to a multiline business was $4.76 per month. This differential is considerably less than that in 1985 when the charges were $1.00 and $4.99, respectively. Switched-access revenues are federally tariffed charges assessed to interexchange carriers for interstate access to LEC facilities.

The primary entrants in network access services business are CAPs who have been nibbling away at the profit margins of the LECs, derived mainly from the commercial market, by facilitating *bypass*.[7] There are two forms of bypass: Customer bypass involves the leasing of a private line from the LEC and connecting it to an interexchange carrier, or another location of the consumer's businesses, thereby avoiding the switched-access services provided by the LEC. Facility bypass involves the installation of a facility that is connected to an interexchange carrier or to another location of the consumer's businesses. The installed facility is usually a high-capacity private fiber-optic "ring," private microwave radio, or via satellite, and is operated by a CAP or by the consumers themselves.

CAPs have been in operation since the mid-1980s and are concentrated in major metropolitan areas providing bypass services to large-volume commercial users. The leading CAPs are Metropolitan Fiber Systems and Teleport Communications Group.[8] Over the years, regulatory agencies across the country have been steadily dismantling barriers to entry in the access services market. A key decision was made in September 1992, when the FCC Expanded Interconnection Order required LECs to allow CAPs to physically co-locate in LEC central offices and interconnect to the LEC network for special access services. Previously, CAPs would have had to physically connect consumers to their own network; with co-location they can directly connect to the entire universe of LEC customers by routing their private traffic through the LEC central office and eventually connecting to their own network or to the interexchange carriers point of presence. By offering co-location to the CAPs, the regulators have virtually opened the door to entrants in the local access business, even though many details need to be worked out and co-location tariffs need to be set.

The CAPs' primary interest is in the revenues to be earned from special-access and private-line services. Special-access charges are for nonswitched private-line access. Private lines provide a dedicated point-to-point connection for the sole use of a single party. These lines are usually available in varying capacity increments ranging from DS-1 circuits that carry 24 voice channels to DS-3 circuits that can carry up to 672 voice channels. These capabilities make such lines ideal for multiline commercial consumers.

Even though CAPs have attracted a lot of attention primarily because of their focus on the lucrative segment of the local telephone business—the high-volume commercial customer—the CAPs themselves have not enjoyed high profit margins. Their business is extremely capital intensive and requires the laying of fiber in congested (and hence, expensive) downtown areas. The fiber ring that encircles an area must then be physically connected to each office building it serves. In fact, several CAPs have had severe liquidity problems and often had to be acquired or merged with other providers (e.g., ICC in Washington, DC and NEDD in Boston).

A key observation that should be made is that the CAPs have attracted customers not on price alone; in fact, frequently, their services have been more expensive than those of the LEC. Service quality and provisioning time coupled with higher reliability and route diversity have been the primary selling points of the CAPs' high-capacity fiber rings. The LECs have responded by making quality improvements themselves. The quality improvements have been taking place simultaneously with the improved services being provided by CAPs; for example, NYNEX, whose major metropolitan areas have been prime targets for CAPs, has responded by introducing high-speed private lines that can provide bandwidth on demand that can be obtained by a customer within an hour. On the other hand, some LECs have made quality enhancing investments in anticipation of such investments by CAPs; for example, Southwestern Bell (whose territory has not yet been subjected to the intensity of competition as has been the case with NYNEX) has built self-healing fiber rings in its major metropolitan areas in advance.

Finally, it must be noted that the degree of substitutability between the package of services offered by an incumbent LEC and a CAP varies significantly. In general, the CAP specializes in providing high-speed private lines and rapid connection of such lines, whereas the LEC has its established set of services and maintenance record and reputation to offer its customers. Given the federal collocation order and with NYNEX's bandwidth-on-demand service or Southwestern Bell's preemptive construction of fiber-optic rings in its major metropolitan areas, the services of several LECs and CAPs are being brought closer together; hence, there is pressure on both the incumbent and the entrant to distinguish their product from that of their rival in other ways. A precedent for such differentiation exists in the interexchange telecommunications industry, in which MCI promotes novel packages such as the Friends and Family program (due to its lead in billing software capabilities), U.S. Sprint's customers have the vital ability to hear pins drop across the globe (due to its lead in fiber optics), and AT&T claims that its services are different for no particular reason other than the fact that they are provided by AT&T.

2.2 The Regulatory Climate

The traditional form of regulation in local telecommunications is rate-of-return (ROR) regulation, which allows the regulated firm to earn a "reasonable" rate of return on its rate base or investment in the telecommunications network within a particular state. The state's PUC determines what is included in the rate base, the depreciation rate, and the acceptable return on the rate base. The allowed ROR is generally an average of the costs of debt and equity weighted by the relative proportions of debt and equity, usually measured at book value (see Phillips, 1988).[9] Any earnings in excess of the allowed ROR must be rebated to rate payers.

ROR regulation is believed to provide the LEC with little incentive to innovate and promote efficiency. For this reason, recently over 65% of states have adopted some form of "incentive" regulation. Each state has its own unique incentive plan that includes combinations of price caps, profits sharing plans, and some partial deregulation of telecommunications services that are deemed to be competitive by the PUCs.

Typically, price cap regulation rather than profits sharing plans provides the best incentives, provided they are implemented properly. Under a pure price cap plan only the price is regulated, and there is no limit set for the maximum that the LEC can earn. These plans commonly adjust for inflation less a specified productivity factor on an annual basis. Typically, the LEC has some flexibility in the sense that several services are grouped together in a basket and only the composite price for the basket is set by the regulator. Even though relatively few states currently employ price cap regulation, it is clearly the wave of the future, especially as competition heats up. For this reason, we use a modified form of price caps in our model of regulation. That is, we assume that the regulator sets up a binding price cap for the regulated firm's service.

With regard to entry policy, currently both the U.S. Senate and House of Representatives are considering legislation that is aimed at fostering competition in local exchange and exchange access services.[10] This federal initiative is in addition to the FCC's Expanded Interconnection Order mentioned earlier. At the state level, by 1993, nine states have already opened their local exchange markets to competition, and 14 additional states have done so on a partial basis.[11] Thus, it seems that regulatory barriers to entry are being rapidly dismantled.

In the following section we consider a simplified version of the model that appears in Chakravorti and Spiegel (1994). Although this is admittedly a highly stylized model, it nonetheless captures key aspects of the rich variety of institutional factors associated with entry in the network access business, mentioned in the preceding paragraphs.

3. THE MODEL AND RESULTS

3.1 The Basic Model

We consider an industry that is initially a regulated monopoly. A key factor in our analysis is the objective of the regulator; we assume that the latter segments the market into two distinct classes which, for the sake of convenience, we label as *residential* and *commercial*. We postulate that the regulator's objective is characterized by a utility function that depends on the welfare of both classes but assigns differential weights to them. Letting U_0 and U denote, respectively, the utilities of the residential consumer and the commercial consumer, the regulator's utility function is given by:

$$W = U_0 + \alpha_1 U, \quad \text{where } \alpha_1 < 1. \tag{1}$$

The regulator sets the price that the incumbent monopoly can charge (i.e., sets a binding price cap) subject to the constraint that the incumbent monopoly can retain at least a fraction $(1 - \alpha_2)$ of its profits, where α_2 is some exogenously negotiated parameter.

For simplicity, we assume that residential consumers demand a fixed quantity, normalized to one unit, and their utility is given by $U_0 = q - p_0 + \tau$, where q is some constant representing the level of the innate quality of the firm's output net of the cost to the firm of providing it, p_0 is the price in the residential market, and τ is a potential transfer payment. The *net price* paid by residential consumers is given by $p_0 - \tau$. We assume that the regulator sets $p_0 = q$; hence, the residential market attracts no entrants. Given this simplification, the utility of residential consumers is captured entirely by the amount of transfer payments they receive, that is, $U_0 = \tau$.

Next, we turn to the source of the transfer payments: the commercial market. The commercial consumers' utility function is different from that of residential consumers. Commercial consumers care not only about the basic service and its quality, q, but also about the extent to which the firm makes quality-enhancing investments, denoted q_r. Examples of such investments in the case of local telecommunications include fiber deployment, provision of private lines that can carry multiple voice channels for high-volume users, enhanced software and personnel capabilities for increased reliability, repair records, and route diversity. The demand for the commercial service as a function of quality-enhancing investment, q_r, and the price in the commercial market, p_r, is:

$$\tilde{x}_r(q_r, p_r) = q + q_r - p_r. \tag{2}$$

The associated indirect utility of commercial consumers is:

$$\tilde{U}(q_r,p_r) = \frac{(q + q_r - p_r)^2}{2}. \tag{3}$$

Assuming that the cost of investing in quality level q_r is $q_r^2/2$, and assuming without a loss of generality that the regulated commercial service can be provided at zero marginal cost, the regulated firm's profits in the commercial market are:

$$\tilde{\pi}(q_r,p_r) = p_r\tilde{x}_r(q_r,p_r) - \frac{q_r^2}{2}. \tag{4}$$

Because the regulator is constrained to let the firm retain at least a proportion $(1 - \alpha_2)$ of its profits, and because $U_0 = \tau$ and $U = \tilde{U}(q_r,p_r)$, his maximization problem is:

$$\underset{p_r,\tau}{Max}\, W = \tau + \alpha_1 \tilde{U}(q_r,p_r), \quad s.t.\ \tau \le \alpha_2\tilde{\pi}(q_r,p_r). \tag{5}$$

It is easy to verify that at the optimum the constraint will be binding. Letting $\alpha \equiv \alpha_1/\alpha_2$, the regulator's problem can therefore be written as:

$$\underset{p_r}{Max}\, W = \alpha_2\, [\tilde{\pi}(q_r,p_r) + \alpha\tilde{U}(q_r,p_r)]. \tag{6}$$

The parameter α captures the regulator's marginal rate of substitution between the welfare of commercial consumers and the firm's profits. We assume that $\alpha < 1$. This reflects the assumption that the marginal utility to the regulator of a unit of profits (used to generate cross-subsidies to residential consumers) is higher than the marginal utility of a unit increment in the welfare of commercial consumers.[12]

3.2 The Monopoly Case

In this subsection we consider the case in which the regulator protects the monopoly position of the regulated firm by blocking entry. Throughout, we assume that investments in enhanced quality cannot be regulated directly, primarily because the costs of verifying the myriad quality-enhancing investments by the firm are too high from a third party's (such as a court's) standpoint, and hence cannot be credibly enforced. We therefore model the strategic interaction between the regulator and the regulated firm as a two-stage game in which the firm makes quality-enhancing investments in Stage 1, and the regulator sets prices in Stage 2. Given the firm's investment in quality and the regulated prices, consumers choose their level of final purchases. The assumption that regulated prices are set after the firm has already invested reflects the fact that adjustments of regulated prices are typically made on a much more frequent basis than firms' investments. The latter, therefore, can be viewed as a long-term decision, whereas the former can be viewed as a short-term decision. These assumptions regarding the sequencing of events also capture the lack of regulatory commitment to prices that characterize the regulatory framework in the United States.

Given our framework, we examine the subgame perfect-equilibrium outcome of the two-stage game. That is, we solve the game backwards by assuming that at each stage players choose their strategies optimally given the history of the game and their correct expectation regarding the outcome of subsequent stages. Note that we do not need to solve directly for τ, because once p_r is set it determines the regulated firm's profits in the commercial market and hence τ, which is a fraction $(1 - \alpha_2)$ of profits.

In Stage 2 of the game, the regulator sets the regulated price for the commercial service. Solving the regulator's maximizing problem, the pricing strategy of the regulator as a function of the quality of the commercial service is:

$$p_r{}^M(q_r) = \frac{(1 - \alpha)(q + q_r)}{2 - \alpha}, \tag{7}$$

where the superscript M is used to index the case of a monopolistic industry.

Anticipating $p_r{}^M(q_r)$, the regulated firm chooses in Stage 1 the level of quality-enhancing investment, with the objective of maximizing its profits. The resulting investment level is:

$$q_r{}^M = \frac{2(1 - \alpha)q}{H}, \tag{8}$$

where $H \equiv 1 + (1 - \alpha)^2 > 0$.

Given the equilibrium level of quality-enhancing investment, the equilibrium regulated price and transfer payment made to the residential market are given as follows:

$$p_r{}^M = \frac{(1 - \alpha)(2 - \alpha)q}{H}; \quad \tau^M = \frac{\alpha_2(1 - \alpha)q^2}{H}. \tag{9}$$

This outcome can be compared with an outcome that would have been realized if in Stage 2 of the game the regulator could control not only the price but also the level of quality-enhancing investment that the firm makes. We refer to this as the complete monitoring outcome. The quality-enhancing investment, regulated price, and transfer payment made to the residential market that emerge from the decisions made in the complete monitoring outcome are given as follows:

$$q_r^* = \frac{q}{1 - \alpha}; \quad p_r^* = q; \quad \tau^* = \frac{\alpha_2 q^2}{2(1 - \alpha)}. \tag{10}$$

These outcomes yield the following insight:

Proposition 1: For all $\alpha \equiv \alpha_2/\alpha_1 > 0$, *relative to the complete monitoring outcome, the regulated firm underinvests in quality-enhancing technologies;*

moreover, prices in the commercial market and cross-subsidies to the residential market are lower.

Proposition 1 has simple intuitive foundations. The regulator cares about the impact of increased quality on the utility of commercial consumers and the impact of quality on profits in the commercial market. The regulator's interest in profits is motivated by the fact that these profits constitute the tax base from which transfers to the residential market are made. The firm, on the other hand, is simply concerned about the second of the two effects, that is, on profits; therefore, it underinvests. As a result both the regulated commercial price and the transfer are lower than they would be in the complete monitoring outcome.

3.3 The Case of Entry

We postulate that the underinvestment in quality that we have identified in the previous subsection provides the regulator with the motivation to exercise the lever that it does have control over: whether or not to permit entry. A priori, one would expect that the presence of competitors would induce the incumbent firm to attract consumers by investing in quality. However, the regulator must weigh this benefit of competition against the potential cost due to the erosion of the regulated firm's profits, which in turn erodes the transfers made to the residential market. This trade-off is further complicated by the possibility that the entrants' product may be differentiated from that of the regulated firm.

An important element in the market for local exchange is the "horizontal" differentiation between the services of different providers.[13] Consider the case of access services with NYNEX, the regional Bell company, and a competitive access provider such as Metropolitan Fiber Systems (MFS). On the one hand, some types of consumers place a high value on the product offered by MFS—a high-speed T1 line within 24 hours, but without the reputation of a long service and repair record—whereas other types place higher value on the longer record of NYNEX, even though their ability to provide high-speed lines within a short time frame is currently limited.

To capture these aspects of entry, we modify the basic model as follows. We assume that the potential entrants constitute a competitive fringe, that is, they are identical to each other (in terms of costs, quality investments, etc.) and play a Bertrand pricing game against each other. This assumption simplifies the analysis considerably. It implies that when entry takes place, entrants will charge a price equal to marginal cost and will therefore have no market power. In the case of the local exchange, this assumption can be justified on the ground that entrants are quite small relative to the size of the market and hence do not have significant market power. In what follows we can therefore focus on a model in which there is a single unregulated firm that considers entry. In addition, we assume that the quality of the service that the entrant can offer is chosen by nature; in other words, the entrant is limited by the technological capabilities available exogenously (such as a fiber-optic ring together with co-location opportunities in the incumbent's central office) and cannot expand its quality offerings. Moreover, it must be noted that because the regulated price in the residential market p_0 is set equal to marginal cost c_0, entry, if it occurs, would be a factor only in the commercial market.

To distinguish the relevant variables of the regulated firm from that of the unregulated entrant, we use the subscripts r and u throughout. Because the entrant charges a price equal to marginal cost, $p_u = c_u$. Assume that if entry occurs, commercial consumers have a quadratic utility function which gives rise to the following demand system:

$$x_r(q_r,p_r) = \left[\frac{q + q_r - p_r - \gamma(q_u - c_u)}{1 - \gamma^2}\right]^+ ; \quad x_u(q_r,p_r) = \left[\frac{q_u - c_u - \gamma(q + q_r - p_r)}{1 - \gamma^2}\right]^+ , \tag{11}$$

where $[\,\cdot\,]^+ \equiv \text{Max}\,\{\,\cdot\,,0\}$, and $\gamma < 1$. The associated indirect utility function of commercial consumers is given by:

$$U(q_r,p_r) = \frac{(q + q_r - p_r)^2 - 2\gamma(q + q_r - p_r)(q_u - c_u) + (q_u - c_u)^2}{2(1 - \gamma^2)}. \tag{12}$$

The parameter γ plays a crucial role in determining the degree of (horizontal) differentiation between the services of the two firms. As γ approaches 1, the degree of differentiation decreases and the two services become closer substitutes except for differences in their quality. Note that when $\gamma = 0$, $x_r(q_r,p_r) = \tilde{x}(q_r,p_r)$. Thus, the demand system specified in Equation 11 extends the demand system specified earlier to the case in which the two services are (at least to some degree) substitutes.

Given $x_r(q_r,p_r)$, the regulated firm's profits are:

$$\pi(q_r,p_r) = p_r x_r(q_r,p_r) - \frac{q_r^2}{2}. \tag{13}$$

We now add to the sequence of events considered in the previous subsection a new stage, Stage 0. In this stage, nature first selects a quality level for the entrant, q_u, and then, after observing q, the regulator decides whether or not to permit entry. The rest of the game (i.e., Stages 1 and 2) is without change. In keeping with the logic of backward induction, we begin with the Stage 2 and solve the regulator's pricing problem as a function of the choices made in earlier stages. Subsequently, we proceed to Stage 1 and solve for the regulated firm's investment problem as a function of the regulator's entry decision made in Stage 0 and assuming that the firm correctly anticipates the regulator's pricing strategy in Stage 2. Finally, we solve for the regulator's entry decision in Stage 0, given the entrant's quality and assuming that the regulator correctly anticipates the regulated firm's investment in enhanced quality in Stage 1.

Stage 2. Replacing $\tilde{U}(q_r,p_r)$ with $U(q_r,p_r)$ and $\tilde{\pi}(q_r,p_r)$ with $\pi(q_r,p_r)$ in the regulator's maximization problem given by Equation 6, and solving for p_r, we have (using the superscript E to denote entry):

$$p_r^E(q_r) = p_r^M(q_r) - \frac{\gamma(1-\alpha)(q_u - c_u)}{2-\alpha}, \tag{14}$$

where $p_r^M(q_r)$ is given by Equation 7. Observe that the regulator's interest in the commercial market price is driven by two factors: A reduction in price increases the welfare of commercial consumers but diminishes the regulated firm's profits and consequently the cross-subsidy to residential consumers. A priori, one cannot be sure how the regulator would respond to the competitive pressure that the entrant exerts. Equation 14 yields a precise answer to this question: The regulated price is lower in the presence of an entrant.

Several key insights are obtained by examining the formula for the price reduction. First, the extent of reduction increases as the regulated and unregulated services become closer substitutes; this simply reflects the intuition that the intensity of competitive pressures are eased as the outputs of the two firms are increasingly differentiated.

Second, the extent of the reduction is greater as the quality of the unregulated firm increases. This is consistent with the observation that if the demand for the regulated service decreases with the quality of the unregulated service (which in turn has a negative effect on the regulated firm's profits and consequently on transfers to the residential market), then the regulated price must be lowered further to bolster demand and improve on the welfare of commercial consumers and thereby offset the decrease in the regulator's utility due to the potential reduction of transfers. The full impact of a higher quality of the unregulated service on the regulated price is not entirely obvious, however, because in the chain of consequences just described we did not factor in the change in the regulated firm's choice of enhanced quality in response to a higher quality of its competitor's service.

Stage 1. In Stage 1 of the game, the regulated firm's profits in the commercial market may be written as $\pi(q_r) \equiv \pi(q_r,p_r^E(q_r))$. Maximizing this expression with respect to q_r, we have:

$$q_r^E = \left[\frac{2(1-\alpha)[q - \gamma(q_u - c_u)]}{H - \gamma^2(2-\alpha)^2} \right]^+. \tag{15}$$

The second-order condition for maximization requires that the denominator of this expression be positive, that is, $\gamma < \tilde{\gamma} \equiv \sqrt{H}/(2-\alpha)$. This condition implies that the regulated service has to be sufficiently differentiated from the unregulated one. We therefore restrict attention to parameter values such that $\gamma < \tilde{\gamma}$.[14] Equation 15 indicates that in equilibrium the regulated firm invests less in enhanced quality as the quality of the unregulated service increases. In fact, when $q_u \geq c_u + q/\gamma$, the regulated firm stops investing in enhanced quality altogether. Also note that $q_r^E > q_r^M$ if and only if $q_u < c_u + \gamma(2 - \alpha)^2/H$. When this condition fails, entry will lead to less investment in enhanced quality, not more as is often assumed by policymakers.

Assuming that $q_r^E > 0$ and substituting from Equation 15 into 14, the equilibrium regulated price in the commercial market is:

$$p_r^E \equiv p_r^E(q_r^E) = \frac{(1 - \gamma^2)(1 - \alpha)(2 - \alpha)[q - \gamma(q_u - c_u)]}{H - \gamma^2(1 - \alpha)(2 - \alpha)}. \tag{16}$$

Given q_r^E and p_r^E, the equilibrium level of subsidy to residential consumers is:

$$\tau^E = a_2\,\pi(q_r^E, p_r^E) = \frac{\alpha_2(1 - \alpha)[q - \gamma(q_u - c_u)]^2}{H - \gamma^2(2 - \alpha)^2}. \tag{17}$$

Similarly, the outputs of the regulated and unregulated firms in the commercial market are:

$$x_r^E \equiv x_r(q_r^E, p_r^E) = \frac{(2 - \alpha)q_r^E}{2(1 - \alpha)}, \tag{18}$$

and

$$x_u^E \equiv x_u(q_r^E, p_r^E) = [q_u - c_u - \gamma x_r^E]^+. \tag{19}$$

Notice that if $q_r^E = 0$, then $\tau^E = x_r^E = 0$. In other words, if the quality of the unregulated service, q_u, exceeds the threshold, $c_u + q/\gamma$, the regulated firm exits the commercial market. In this case, the commercial market is served by an unregulated monopoly providing an output $q_u - c_u$, and residential consumers receive no cross-subsidies. On the other hand, when $q_u \leq c_u + \gamma x_r^E$, the unregulated firm cannot penetrate the commercial market because $x_u^E = 0$.

We summarize this discussion in the following proposition:

Proposition 2: *There are three different cases to consider depending on the size of q_u:*

(i)   *For all $q_u \leq c_u + \gamma x_r^E$ no entry occurs.*

(ii)   *For all $c_u + \gamma x_r^E < q_u < c_u + q/\gamma$, the market is served by both firms.*

(iii) *For $q_u \geq c_u + q/\gamma$, the regulated firm exits the commercial market.*

Proposition 2 demonstrates the clear link between the entrant's quality and its ability to penetrate the commercial market and affect the decisions of the regulated firm. As the entrant's quality increases, its impact increases: At low-quality levels, the entrant is kept out of the market.[15] At intermediate quality levels, entry takes place, and for high-quality levels, it is the regulated firm that is forced out of the commercial market, and the industry becomes an unregulated monopoly. Furthermore, observe that $\partial(\gamma x_r^E)/\partial\gamma > 0$ and $\partial(q/\gamma)/\partial\gamma < 0$. Hence, as the regulated and unregulated products become closer substitutes: (a) the regulated firm expands its output, so the unregulated firm needs to offer higher quality to penetrate the market; and (b) the unregulated firm is more likely to monopolize the commercial market once it enters.

Stage 0. Now consider the regulator's decision on whether or not to permit entry. The regulator's decision is based on a direct comparison between the value of his or her payoff with and without entry, for each one of the cases discussed in Proposition 2. As it turns out, this comparison depends in a complex way on the parameters of the model. A detailed analysis of the regulator's decision is given in Chakravorti and Spiegel (1994) to which the reader is referred. In what follows, we briefly review some of the main results. To simplify matters, we consider here only cases in which $q_u > c_u + \gamma x_r^E$, that is, cases in which entry occurs if it is allowed.

To decide whether or not to permit entry, the regulator must perform the following cost-benefit evaluation: The cost of permitting entry is the loss of cross-subsidies to residential consumers due to erosion of monopoly profits in the commercial market. The benefit of permitting entry is the increase in welfare of commercial consumers due to changes in quality, price, and greater product variety.[16]

We begin by considering the case in which the entrant's quality is so high that he drives the regulated firm out of the commercial market. Intuitively, it is clear that the regulator would permit entry whenever the quality of the entrant's service is sufficiently high to ensure that the benefit to commercial consumers from having a superior service outweighs the loss to residential consumers from having to concede their cross-subsidies. Note, however, that this argument depends on the relative weights that the regulator assigns to the welfare of each group of consumers. For instance, if the regulator cares only about residential consumers ($\alpha = 0$), then he completely ignores the benefits of entry to commercial consumers and would therefore always block entry. On the other hand, if the regulator cares about the welfare of commercial consumers ($\alpha > 0$), then he would permit entry if the quality of the unregulated service is sufficiently high. To illustrate these points, we therefore consider the following example. Let $\gamma = 1/2$, $q = 10$, and $q_u - c_u = 20 + \delta$, where $\delta \geq 0$ (note that in order to drive the regulated firm out of the market, the unregulated firm must have $q_u - c_u \geq q/\gamma = 20$). Given these values, a tedious but straightforward calculation reveals that the regulator will permit entry if and only if:

$$\alpha(20 + \delta)^2(2 - 2\alpha + \alpha^2)^2 > 100[4(1 - \alpha) + \alpha^2(2 - \alpha)]. \qquad (20)$$

Note that when $\alpha = 0$, the left side vanishes so the condition fails. On the other hand, when $\alpha = 1$, the left side equals $(20 + \delta)^2$, whereas the right side equals 100, so the condition holds. Moreover, note that the left side of the expression increases with δ. From this we conclude that the regulator will allow entry only if α and δ are relatively large (i.e., the regulator places a relatively high weight on the welfare of commercial consumers, and the net quality of the unregulated service is relatively high).

> **Proposition 3:** *Suppose that the entrant's quality is so high that if entry is allowed, he drives the regulated firm out of the commercial market, and let $\gamma = 1/2$, $q = 10$, and $q_u - c_u = 20 + \delta$, where $\delta \geq 0$. Then, the regulator will allow entry if and only if α and δ are sufficiently high to ensure that condition (20) is satisfied.*

Next, consider the case in which q_u is intermediate so that both firms are active in the commercial market. Recall that entry into the commercial market improves on the welfare of commercial consumers but makes residential consumers worse off. Thus, it is intuitively clear that the regulator would block entry if he places a small weight on the welfare of commercial consumers (i.e., α is small). On the other hand, when the regulator places a high weight on the welfare of commercial consumers, the reverse holds. The reason for this is that as α approaches 1, the regulator's objective approaches the maximization of social welfare. Consequently, the regulator will set the regulated price equal to marginal cost, leaving the firm with zero profits and, hence, no cross-subsidy is being generated (note from Equation 17 that $\tau^E \to 0$ as $\alpha \to 1$). Thus, in this case, only the welfare of commercial consumers matters. But, because entry provides commercial users with more variety, it makes them, and therefore the regulator, better off. Similarly, when the two services are poor substitutes, the entrant has a negligible effect on the regulated firm's profits and consequently the cross-subsidy to residential consumers (note from Equations 17 and 9 that $\tau^E \to \tau^M$ as $\gamma \to 0$). But, because entry improves on the welfare of commercial users by providing them with more variety, the regulator will allow it. Hence,

> **Proposition 4:** *Suppose that q_u is intermediate in the sense that both the regulated and unregulated firms are active in the commercial market. Then the regulator will allow entry if α is close to 1, or γ is close to 0. On the other hand, the regulator will block entry if α is close to 0.*

The conclusion from this analysis can be summarized as follows: The regulator is more inclined to permit entry as α—the measure of relative weight attached to the welfare of commercial consumers—increases,[17] as the quality of the unregulated service increases, and as the degree of differentiation between the regulated and the unregulated products increases.

4. CONCLUSION

Our central conclusion is that the question of whether or not there is a dismantling of regulatory entry barriers in protected markets, such as that for network access services in local telecommunications, hinges on three factors: (a) the extent to which the regulator cares for residential consumers versus his concern for commercial consumers, (b) the innate quality of the entrant's service, and (c) the extent to which the entrant offers a service that is not a close substitute of the one available from the incumbent firm.

Although the first two factors appear to be intuitively obvious, it is quite interesting to note the key role that is played by the third factor—product differentiation between entrant and incumbent. The impact of such differentiation can be summarized as follows: Greater differentiation (a) expands the product space; (b) reduces the erosion of the regulated firm's profits and thereby reduces the decline in cross-subsidies; (c) increases the probability that the unregulated firm will enter by lowering the minimal quality investment needed to penetrate the market; (d) conditional on entry, decreases the probability that the unregulated firm will be a monopoly by raising the minimal quality threshold which the unregulated firm must attain in order to induce the regulated firm to exit; and (e) reduces the decline in the regulated post-entry price in the commercial market.

These general conclusions provide a valuable guide both to strategic planners in incumbent firms as well as to policymakers. The message to the former is that if the firm's objective is to lower the probability that regulatory entry barriers are lifted, then the firm should try to offer as close a substitute for the entrant's service as possible. This strategy is not without risks, however, because our analysis shows that as the two services become closer substitutes, the entrant is more likely to drive the regulated firm out of the market if entry occurs after all. The message to policymakers is that in considering whether or not to permit entry of unregulated firms, one needs to consider not only the quality of the entrants' service, but also how close of a substitute it is for existing services. This consideration is going to affect not only the emerging market structure (i.e., whether there will be one or more providers of services in the market), but also the the regulated firm's investment incentives and its ability to generate cross-subsidies.

ACKNOWLEDGMENTS

The views expressed in this chapter are those of the authors and do not represent the views of Bellcore or any one of its owner companies. We thank Bill Lehr for helpful comments.

ENDNOTES

1. For a historical account of the origins of cross-subsidies in the U.S. telecommunications industry, see Temin (1990).
2. "Potential Impact of Competition on Residential and Rural Telephone Service," USTA Study, July 21, 1993.

3. Source: Table 6 in "Trends in Telephone Services," Industry Analysis Division, FCC, October, 1993.

4. Statement of Reed Hundt, Chairman of the FCC, before the Subcommittee on Telecommunication and Finance, U.S. House of Representatives, on H.R. 3636, the "National Communications Competition and Information Infrastructure Act of 1993," and H.R. 3626, the "Antitrust Reform Act of 1993" and "Communications Reform Act of 1993," January 27, 1994.

5. The RBOCs' access lines in the contiguous United States were 105.7 million out of a total of 137.7 million (source: Table 14, "Trends in Telephone Services," Industry Analysis Division, FCC, October, 1993). The RBOCs reported revenues from local exchange services of $31.2 billion in 1992 as compared with $39.2 for all LECs (source: Tables 2 and 7, "Telecommunication Industry Revenue: TRS Fund Worksheet Data," Common Carrier Bureau, Industry Analysis Division, FCC, March 1994).

6. Source: Table 7, "Telecommunication Industry Revenue: TRS Fund Worksheet Data," Common Carrier Bureau, Industry Analysis Division, FCC, March 1994.

7. In January 1994, MCI announced its own plan to enter the local exchange market via a newly created subsidiary, MCI Metro. In entering the market, MCI can take advantage of its recent acquisition of Western Union, whose underground conduit system runs throughout the downtown areas of most major metropolitan areas.

8. As of 1992, Metropolitan Fiber Systems served 14 major metropolitan areas operated in 12 different states, and Teleport Communications Group served 12 major metropolitan areas in 8 different states (source: "Fiber Deployment Update—End of Year 1992," by Jonathan Kraushaar, Industry Analysis Division—Common Carrier Bureau, FCC, April 1993).

9. In fact, this procedure is followed not just under ROR regulation, but also under price cap regulation because price caps are set on the basis of the firm's cost of capital. For example, the FCC sets price caps on interstate access rates so as to ensure local exchange carriers a rate of return on their investment of 11.25%. Similarly, the FCC has tentatively concluded to establish price caps on cable TV services so as to ensure cable operators a rate of return on their investment of approximately 10%–14%.

10. This legislation includes S. 1822, the Communications Act of 1994, and H.R. 3636, the National Communications Competition and Information Infrastructure Act of 1993.

11. States that allow competition in local exchange include Illinois, Iowa, Michigan, Montana, Nebraska, New York, Oregon, Pennsylvania (nonswitched local), and Washington. States that allow partial competition include California, Colorado (nonswitched local service), Washington, DC, Florida, Maryland (not switched), Massachusetts, Missouri (nonswitched local service), New Jersey (not basic local exchange), North Dakota, Texas (certain nonbasic services), Virginia, West Virginia (not basic local), Wisconsin (nonswitched local), and Wyoming (source: Table 165, "Competition in Local Exchange Service," NARUC, Compilation of Utility Regulatory Policy 1992–1993).

12. This assumption is consistent with Kaserman and Mayo (1994), who argue that residential customers appear to be in a better position to exert political pressure on regulators than commercial customers. Thus, regulators care more about residential customers due to their concern over the political repercussions of infringing on the rights of this group's interests.

13. The difference between horizontal and vertical differentiation is the following. When services are vertically differentiated, all consumers agree on which service is better. In contrast, when services are horizontally differentiated, there is no unanimous agreement between consumers on which service is better: some consumers prefer the attributes of one service more than the other, while other consumers have the reverse preferences.

14. Differentiating $\tilde{\gamma}$ with respect to α reveals that $\delta\tilde{\gamma}/\delta\alpha = \alpha\tilde{\gamma}/H > 0$. Hence, $\tilde{\gamma}$ increases from $1/\sqrt{2}$ when $\alpha = 0$, to 1 when $\alpha = 1$.

15. In our related paper (Chakravorti and Spiegel, 1994) we show that even if the unregulated firm stays out of the market, its presence induces the regulated firm to increase its investment in enhanced quality nevertheless, provided that the quality of the unregulated service is not

too low. The reason for this is that the unregulated firm stays out of the market precisely because the regulated firm invests more than it would otherwise.

16. In Chakravorti and Spiegel (1994), we prove formally that entry always makes commercial consumers better off and residential consumers worse off than they respectively are when entry is blocked. Note that the first result is not obvious a priori because entry may lead to a reduction in the regulated firm's investment in enhanced quality. As it turns out, however, the reduction in the regulated price and the increase in variety are sufficient to compensate commercial consumers for the reduction in quality if it occurs.

17. In other words, the regulator is less inclined to allow entry if he places a relatively high weight on profits (i.e., α is small). This is reminiscent of Weisman (1993), where the regulator can rebate to consumers a share of the regulated firm's profits (rather than use them as a cross-subsidy as in our chapter). Weisman shows as the regulator can rebate to consumers a larger share of profits, he becomes less inclined to permit entry.

REFERENCES

Chakravorti, B. and Y. Spiegel, (1994), "A Political Economy Model of Entry Into Regulated Markets," Mimeo.

Daniel, T., R. Shin, and M. Ward, (1993), "What's Regulators To Do? Responses to Competition in Local Telephone Service," Mimeo.

Kaserman, D., and J. Mayo, (1994), "Cross-Subsidies in Telecommunications: Roadblocks on the Road to More Intelligent Telephone Pricing," *Yale Journal on Regulation*, Vol. 11, pp. 119–147.

Palmer, K., (1992), "A Test for Cross Subsidies in Local Telephone Rates: Do Business Customers Subsidize Residential Customers?" *The Rand Journal of Economics*, Vol. 23, pp. 415–431.

Posner, R., (1971), "Taxation By Regulation," *The Bell Journal of Economics*, Vol. 2, pp. 22–50.

Phillips, C., (1988), *The Regulation of Public Utilities*, 2nd Edition, Arlington, VA: Public Utilities Reports, Inc.

Temin, P., (1990), "Cross Subsidies in the Telephone Network after Divestiture," *Journal of Regulatory Economics*, Vol. 2, pp. 349–362.

Weisman, D., (1993), "Why Less May Be More Under Price-Cap Regulation," Mimeo.

Chapter 3

Dynamic Effects of Regulation on Exchange Carrier Incentives

Neal Stolleman
Bellcore

I. INTRODUCTION AND OVERVIEW

The purpose of this chapter is to examine the effect of alternative forms of regulation on an exchange carrier's pricing and investment decisions, where investment is analyzed in terms of capacity expansion and in terms of improvements in infrastructure quality. Two forms of regulation are considered: rate of return and price cap regulation. It is becoming increasingly important to analyze the differential effects of these policies as exchange carriers expand their operations in both regulated and unregulated markets, further exacerbating the problems associated with implementing cost allocation rules. These rules serve to apportion the costs of resources used in the joint production of multiple services, which influences the rates charged for these services and thereby affect the firm's investment decisions. Hence, the choice of regulatory policy has a direct effect on the efficiency with which the telecommunications infrastructure evolves, as well as important implications for the quality of that infrastructure.

The analytical model presented here offers a stylized version of an exchange carrier in which the firm sells both a regulated, basic service[1] and an unregulated, enhanced service. The basic service is sold directly to retail customers in a regulated market and as an input to value-added resellers (VARs). The VARs combine the basic service with additional resources to produce an enhanced service that is sold in an unregulated, competitive retail market. The exchange carrier may be affiliated directly with one of the VARs. Moreover, I assume that the exchange carrier is subject to Open Network Architecture (ONA) constraints,[2] which preclude the exchange carrier from price discriminating when it sells the basic service. This means that the price for the basic service will be the same whether it is sold to consumers directly in the regulated retail market or as an input for enhanced services in the wholesale market to VARs. Moreover, the exchange carrier cannot charge different prices to affiliated and unaffiliated VARs. The VARs compete in a competitive retail market.

In addition to the exchange carrier, I assume that there are alternative suppliers of underlying infrastructure, including interexchange carriers (IXCs), cable companies (CATV), and competitive access providers (CAPs) such as Teleport and Metropolitan Fiber Systems from whom the VARs may purchase the basic services that are used as an input to produce the enhanced service. Some of these alternative infrastructure suppliers also may participate as VARs. Thus, the exchange carrier participates in the enhanced market both directly via an affiliated VAR and indirectly via sales of its basic services in the wholesale market to other VARs.

Recent trends toward deregulation that are encouraging the unbundling of access to local exchange networks and the proliferation of new retail operations offering wireless and interactive video services, which are overlaid on the wireline infrastructure, are leading us toward this type of industry structure.

The vertically integrated exchange carrier is assumed to use variable resources (e.g., labor, materials, energy, etc.) and network capital to produce its basic and enhanced services. This chapter differs from similar analyses by distinguishing between two types of network capital investments. The firm can invest in both expanding its physical, or nominal, capital (K_N) as well as the productivity of its capital (B). These latter types of investments, which are not traditionally included in the computation of the firm's rate base, may be thought of as investments to enhance the quality of network capital or as investments in innovation. Cost allocation rules determine how costs are allocated between basic and enhanced services.

The model that is presented and analyzed in the balance of the chapter yields five main conclusions, or recommendations, as follows:

1. *Policymakers should make sure that the price cap formula is relatively insensitive to fluctuations in total sales of the basic service.* The price cap mechanism specifies how the maximum price that may be charged is periodically adjusted. If increases in output lead to steep declines in the price cap, then a spiral of decreasing prices and increasing output may result. Or, going in the other direction, contracting output may lead to a spiral of increasing prices until the cap no longer constrains the firm's behavior. This kind of instability should be avoided.

2. *Policymakers need to make sure that the price cap formula does not induce the carrier to defer investments that enhance the quality of network capital.* Downward adjustments in the price cap to reflect productivity gains reduce incentives to invest in improving the quality of network capital. There are at least three

dynamic price cap adjustment effects that may discourage such investments. First, increases in near-term, quality-enhancing investments that are treated as ordinary operating cost increases will appear to make the price cap more binding. Second, in the longer term, these investments will make the installed stock of capital more productive, leading to a decline in the price cap over time. And, third, increased productivity of capital may lead to a decline in the expected real price of capital over time, lowering incentives to undertake current investment. If policymakers embed these sorts of adjustments in the price cap formula, then current expenditure on welfare-enhancing investments will be discouraged. Current investment may also be deferred if regulatory policy changes that are more conducive to such investments are expected to occur sometime in the future.

3. *Policymakers can encourage increased investment in both physical capital and quality-enhancements via suitable adjustments to the price cap formula.* This is simply the flip side of point (2). Current period adjustments in the price cap formula that provide allowances for increases in the rate of investment and for expenditures that are likely to increase infrastructure quality (e.g., R&D) would strengthen incentives in the proper direction. It would also counteract incentives to defer decisions caused by expectations mentioned in (2).

4. *Price cap regulation produces more efficient behavior than either pure Rate of Return (RoR) or a hybrid of price cap and RoR regulation.* Both types of regulation distort the exchange carrier's investment decisions. Although both policies distort pricing behavior and hence may distort the level of resources devoted to serving regulated and unregulated markets, price cap regulation is more likely to result in an efficient path for the evolution of network capital. Incorporating the adjustments noted earlier and eliminating the confusion introduced by divergent state and federal policies will improve the efficiency of price cap regulation. Achieving these goals, however, is likely to be politically quite difficult.

5. *In the long run, an increased rate of investment in the near term will result in increased productivity that will feed through the price cap formula leading to a stable and possibly declining price cap.* In other words, positive adjustments in the price cap to provide investment incentives of the sort discussed in (3) will eventually lead to a stable and possibly declining price cap as the stock of network capital accumulates with higher average levels of productivity.

These five conclusions are based on an analysis of an abstract mathematical model that represents an exchange carrier's pricing and investment behavior as a constrained, dynamic optimization problem. The carrier seeks to maximize the present value of its profit streams over its planning horizon subject to the constraints imposed by price cap and RoR regulation. The exchange carrier sets the price for its basic service, chooses rates of investment to augment physical capital and the quality of that capital, and chooses the level of sales of the enhanced service. These decisions are reevaluated at each point in time along the firm's planning horizon, conditional on the consequences of last period's decisions. I assume that in the absence of regulatory distortions, the exchange carrier's pricing, investment, and innovation decisions would be efficient in the sense that they maximize the present value of the firm's profit streams. Having defined my notation and presented the model, I proceed to analyze the first order necessary conditions that help identify an optimal solution. The effects of regulatory distortions on pricing, output, and investment behavior are deduced via consideration of selected terms in these first-order conditions. Although the basic framework is quite general, important assumptions that help specialize my model (e.g., regarding regulatory behavior) are discussed along with notational conventions employed.

The balance of this chapter is organized into four sections. Section II describes the model's notation and structure as well as important assumptions (e.g., regarding regulatory behavior and structure of markets). Section III analyzes the first-order conditions, whereas Section IV suggests policy innovations that may alleviate the regulatory distortions analyzed in the preceding section. Section V offers a concluding summary.

II. MODEL: STRUCTURE, NOTATION, AND ASSUMPTIONS

The generic form of the constrained, dynamic optimization model used in this analysis is as follows:

$$\max_{(P_R, I_N, I_B, Q_X)} \int e^{-rt} \left\{ Profit_t - \lambda_{1,t}[RoR\ Constraint_t] + \lambda_{2,t}[Price\ Cap\ Constraint_t] \right\} dt \quad (1)$$

Equation 1 shows that at each point in time the exchange carrier chooses values for the price of the basic service (P_R), rates of gross investment in nominal capital (I_N) and quality (I_B), and sales of the enhanced service (Q_X) so as to maximize the present value of future profit streams. These decisions are made conditional on existing technological and market conditions, the cumulative levels of network capital and quality and, importantly, the prevailing set of regulatory constraints. If there were no binding regulatory constraints impinging on exchange carrier decisions, then both of the Lagrange multipliers $\lambda_{1,t}$ and $\lambda_{2,t}$ would be equal to zero at each point in time.[3] By constraining one or the other of these coefficients equal to zero, it is possible to separately analyze the distortions each form of regulation introduces into the firm's decision making as well as distortions created when both forms of regulation exist together in a hybrid system.

Specification of the model and its analysis requires defining a large number of variables and functional relationships. These are grouped into relevant categories (by type of variable) into Tables 3.1 through 3.7. When the variables and

TABLE 3.1
Output Quantity and Price Variables

Q_R	Quantity demanded of the basic service as a function of the regulated price (demand emanates from both retail and wholesale markets)
η_R	Composite (retail and wholesale) price elasticity of demand for basic service
$Q_{R,X}$	Derived demand for the basic service used to produce exchange carrier's enhanced service output, depends on price of basic service and production level of enhanced service
$\rho_{R,X}$	Fraction of basic service output used to produce exchange carrier's enhanced service $= Q_{R,X}/Q_R < 1$
Q_X	Quantity of enhanced service produced by exchange carrier, depends on the intersection of marginal production cost and the competitive equilibrium enhanced service price
P_R	Price of the basic service (also a control variable for the exchange carrier; see Table 3.4)
P_X	Competitive equilibrium price for enhanced service determined by the intersection of the market demand and industry supply curves for the enhanced service (net of value-added resource costs)

TABLE 3.2
Input Quantity and Price Variables

K_N	Quantity of nominal or physical network capital used to produce the basic, regulated service
B	Level of quality of physical network capital due to exchange carrier's state of technical knowledge (i.e., the efficiency or productivity of network capital)
K	Effective amount of network capital $= K_N B$ (i.e., quality adjusted capital)
P_K	Acquisition price per unit of nominal capital
P_K/B	Acquisition price per unit effective network capital
$dLn(P_K)/dt$	The expected rate of change in the acquisition price of nominal capital
$dLn(B)/dt$	The expected rate of change in the quality of nominal capital
$dLn(P_K/B)/dt$	The expected rate of change in the acquisition price of effective capital

TABLE 3.3
Environmental Variables

r	Discount rate
w	Depreciation rate for physical capital
h	Depreciation rate of exchange carrier technical knowledge

TABLE 3.4
Control Variables

P_R	Price of the basic service
I_N	Gross additions to the nominal stock of network capital
I_B	Gross additions to the state of technical knowledge or quality of network capital

TABLE 3.5
Equations of Motion

dK/dt	Net change in the stock of effective network capital
	$= I_N B - wK$
dB/dt	Net change in the level of network quality
	$= I_B - hB$

TABLE 3.6A
Primary Cost Functions

C^R	Expenditure on variable resources producing basic, regulated service
	$= C^R(Q_R, K, I_N)$, with partial derivatives:
$\delta C^R/\delta Q_R > 0$	Marginal cost of increasing production of the basic service
$\delta C^R/\delta K < 0$	Marginal savings in variable resources when effective capital is increased (a substitution effect between capital and labor)
$(1/B)\delta C^R/\delta I_N > 0$	Internal adjustment cost of changing the rate of gross investment in nominal capital
	$\equiv \phi_R$
$dln\phi_R/dt$	Expected rate of change in internal marginal investment adjustment cost
C^B	Expenditure on resources to augment the quality of network capital
	$= C^B(I_B)$, with partial derivative:
$\delta C^B/\delta I_B > 0$	Internal adjustment cost of changing the rate of gross investment in network quality
	$\equiv \phi_B$
$dLn\phi_B/dt$	Expected rate of change in internal marginal quality adjustment cost

TABLE 3.6B
Derived Cost Functions

U_K	User cost of effective network capital (External adjustment cost)
	$= (P_K/B) (r + w - dLn(P_K/B)/dt)$
U_C	User cost of effective network capital (Internal adjustment cost)
	$= (1/B)\delta C^R/\delta I_N (r + w - dLn(\phi_R/B)/dt)$
U_B	User cost of network quality (Internal adjustment cost)
	$= \delta C^B/\delta I_B (r + h - dLn(\phi_B)/dt)$

TABLE 3.7A
Regulatory Design Variables: Rate of Return Variables

S	Maximum rate of return allowed on the regulated base rate
α_K	Fraction of the nominal capital stock assigned to the regulated rate base

TABLE 3.7B
Regulatory Design Variables: Price Cap Variables and Relations

ψ	Maximum allowed price for the basic, regulated service under a stylized price cap system
	$= \psi(Q_R, K, I_N, B, I_B)$, with the partial derivatives:
$\delta\psi/\delta Q_R < 0$	Marginal decrease in maximum allowed price due to an increase in production that raises productivity
$\delta\psi/\delta K < 0$	Marginal decrease in maximum allowed price due to an increase in effective network capital that raises productivity
$(1/B)\delta\psi/\delta I_N > 0$	Marginal increase in maximum allowed price due to an increase in the rate of gross capital investment that drains cash flow and lowers productivity (MIPCA)
	$\equiv \psi_N$
$dLn\psi_N/dt$	Expected rate of change in marginal investment price cap adjustment
$\delta\psi/\delta B < 0$	Marginal decrease in maximum allowed price due to an increase in the quality of network capital that raises productivity
$\delta\psi/\delta I_B > 0$	Marginal increase in maximum allowed price due to an increase in the rate of gross additions to the state of technical knowledge that drains cash flow and lowers productivity (MQPCA)
	$\equiv \psi_B$
$dLn\psi_B/dt$	Expected rate of change in marginal quality price cap adjustment

TABLE 3.7C
Regulatory Constraints

λ_1	Lagrangean multiplier for rate-of-return regulation, a positive number less than 1, showing the change in maximum attainable profit due to a small relaxation of the RoR constraint. If RoR is not a binding constraint, this variable equals zero. The RoR constraint is defined as:
	$P_R Q_R - C^R(Q_R, K, I_N) - C^B(I_B) \leq S\alpha_K P_K K_N$
λ_2	Lagrangean multiplier for price cap regulation, a negative number in the model formulation. The change in maximum attainable profit due to a small relaxation of the price cap constraint is $-\lambda_2 > 0$. If price cap regulation is not a binding constraint, this variable equals zero. The price cap constraint is defined as:
	$P_R \leq \psi(Q_R, K, I_N, B, I_B)$

functional relations described in these tables are inserted into the generic form of Equation 1, the dynamic profit maximization problem is expressed as:

$$\max_{P_R, I_N, I_B, Q_X} \int e^{-rt}\{P_R Q_R - C^R(Q_R, K, I_N) - C^B(I_B) - U_K K$$

$$+ P_X Q_X - P_R Q_{R,X}$$
$$- \lambda_1[P_R Q_R - C^R(Q_R, K, I_N) - C^B(I_B) - S\alpha_K P_K K_N]$$
$$+ \lambda_2[P_R - \psi(Q_R, K, I_N, B, I_B)]\}dt \qquad (2)$$

Equation 2 explicitly shows the revenue and cost components as well as the elements of the regulatory constraints. Revenue is generated from the sale of both basic (P_RQ_R) and enhanced services (P_XQ_X). Costs are generated by the variable resources used to produce the basic service ($C^R(Q_R,K,I_N)$), by resources associated with improving the quality of network capital ($C^B(I_B)$), and by resources associated with the physical stock of network capital (U_KK). The capital related costs are essentially the costs of "holding" the stock of capital and include foregone interest income on the dollar value of the stock, depreciation charges, and the expected rate of change in the real price of capital. There is also an internal transfer price that the exchange carrier pays to itself under ONA pricing assumptions for that portion of the basic service used as an input in enhanced service production.

Equation 2 is maximized by choosing values for the control variables (P_R, I_N,I_B) subject to the two types of regulatory constraints that are explained more fully later. The optimal solution is found as the solution to a system of first-order necessary conditions. These first-order conditions (discussed more fully in section III) identify when the marginal benefits and marginal costs associated with each control variable are balanced in equilibrium. In principle, this optimizing process is repeated at each point in time over the firm's planning horizon, generating trajectories for prices, output, investment flows, and cumulative stocks of capital and quality. Regulatory policy intrudes into the optimizing process by skewing cost–benefit calculations, thus shifting the control variables away from their efficient trajectories. In section III, I analyze the nature of these distortions in more detail, providing a basis for recommendations that are made in section IV.

The first regulatory constraint to consider is rate-of-return regulation, which is expressed as:

$$P_RQ_R - C^R(Q_R,K,I_N) - C^B(I_B) \leq S\alpha_K P_K K_N \qquad (3)$$

The left-hand side (LHS) of Equation 3 is the gross profit realized from regulated operations (regulated revenues less variable resource and innovation related costs). The right-hand side (RHS) is the authorized rate of return (S) applied to the fraction of the capital stock assigned to the regulated rate base, α_K. Policymakers control the maximum allowed gross profit by adjusting S or α_K. Note that unregulated revenues must cover the remaining ($1 - \alpha_K$) of capital costs not assigned to the regulated rate base. Furthermore, I assume that (a) once set by regulators, the value of α_K is independent of the mix of regulated and unregulated services; and (b) all of the variable resource- and innovation-related costs are assigned to regulated operations.[4]

The second regulatory constraint—price cap regulation—is expressed as:

$$P_R \leq \psi(Q_R,K,I_N,B,I_B) \qquad (4)$$

The LHS of Equation 4 is the price charged by the exchange carrier for the basic, regulated service. The RHS is defined as the maximum allowable price for the basic service as set by the price cap formula. The maximum allowable price is, in turn, a function of a set of output, capital, and quality related variables. In this stylized representation, the maximum allowable price varies inversely with the level of the firm's productivity. Hence, the differential effect on the price cap of each variable on the RHS of Equation 4 depends both on how each variable influences the firm's productivity and on how much weight the productivity effect is given in the price cap formula. Increases in output, the effective capital stock, and the quality of capital are assumed to improve productivity and therefore lead to a decline in the price cap. Increases in the rate of gross investment in either physical capital or quality drain resources from current production, lower productivity, and result in a higher price cap. Each of these partial differential effects are described in Table 3.7B. Regulators can give more or less weight to each of these partial differential effects and thereby control each variable's influence on the price cap.

An essential part of the firm's constrained optimization process is that choice values for the control variables (P_R, I_N,I_B) can never lead to violations of the constraints defined in Equations 3 and 4. The sizes of the "multipliers," λ_1 and λ_2, indicate how stringent the regulatory constraints are in terms of influencing the firm's decision making. By setting one or the other multiplier equal to zero, the distortions each form of regulation introduces into the first-order cost–benefit calculations can be ascertained.

The exchange carrier is assumed to be the only entity selling the basic, regulated service directly to retail consumers. It also sells the basic service to value-added resellers for use in the production of the enhanced service. Because of ONA pricing restrictions, the price charged is the same regardless of customer type. The elasticity of demand for the basic service is a composite of retail and wholesale demand elasticities, with the latter presumably larger because of the ability of VARs either to turn toward alternative infrastructure suppliers or to exit the industry. However, the presence of potential competitive pressures in the local distribution network tempers the exchange carrier's decision making even though it is the only actual retail basic service supplier.

The exchange carrier participates in the enhanced services market indirectly by supplying its basic service arrangement to all value-added resellers, including its own affiliated VAR. The market for the enhanced service is assumed to be competitive in the sense that no one firm is large enough to exert influence over the equilibrium price, and each firm chooses its level of sales so as to equate marginal cost with the competitively determined price. Nevertheless, VARs can differ in terms of managerial quality, even if identical technologies are used. Therefore, the industry supply curve for the competitive industry may be upward sloping. Intersection of the industry supply and market demand curves determines the equilibrium price for the enhanced service.

Because the basic, regulated service is used as an input by all VARs in the enhanced service industry, an increase in its price will shift the industry supply curve upward. An increase in the regulated price, therefore, will cause marginal VARs to exit the industry because their internal cost curves will have increased. Alternatively, marginal VARs may turn to alternative suppliers of infrastructure. Equilibrium output of the remaining entities will be lower due to their higher input costs, and a new, higher price for the enhanced service will result. The higher price will curtail market demand by enough to match the reduced industry output. Therefore, both the equilibrium price of the enhanced service and the level of sales of each VAR is partially a function of the price of the basic, regulated service. I examine these market interactions more formally in Section III.

III. FIRST-ORDER CONDITIONS FOR PRICING, INVESTMENT, AND QUALITY DECISIONS

In the following four subsections I analyze the model described earlier. The first three subsections examine the first-order conditions associated with the carrier's choice of a price for its basic service (P_R), the level of investment in physical capital (I_N), and the level of network quality (B), which implies a level of investment in improving network quality. I ignore the choice of Q_X, the level of sales of the enhanced service by the VAR, because this is determined by competitive supply and demand factors that are not directly controllable by the exchange carrier. The last subsection analyzes alternative regulatory frameworks.

III.A. Pricing Decision for the Basic Service

The first-order condition that determines the price charged for the basic, regulated service at each point in time (P_R) is[5]:

$$[P_R - (\delta C^R/\delta Q_R)]/P_R = 1/\eta_R$$
$$+ [1/(1 - \lambda_1)][(Q_X/Q_{R,X})(\delta P_X/\delta P_R) - 1]\rho_{R,X}/\eta_R$$
$$+ [\lambda_2/(1 - \lambda_1)][(\delta\psi/\delta Q_R) - (\delta P_R/\delta Q_R)](1/P_R) \qquad (5)$$

The first line in Equation 5 is the percentage markup of price over the marginal cost of producing the basic service that maximizes the firm's profits. In the absence of regulatory constraints and vertical integration into the enhanced services market, standard economic theory dictates that this markup should be set equal to the the inverse of the price elasticity of demand, $1/\eta_R$, which is another way of expressing the well-known profit maximization condition equating marginal revenue and marginal cost. The next two lines of Equation 5 reflect the added complexity introduced by vertical integration and regulation.

Despite its complexity, Equation 5 yields some intuitively appealing interpretations. For example, because ONA rules prohibit price discrimination among VARs, higher prices for the basic service increase the costs of offering the enhanced service by the affiliated VAR. Therefore, the exchange carrier's incentive to increase its price markup for the basic service will be inhibited the larger the fraction of its basic service used in enhanced service production, $\rho_{R,X}$. On the other hand, an increase in the basic service price leads to an increase in the equilibrium price of the enhanced service. To the extent that the increase in the exchange carrier's internal costs can be recovered by corresponding increases in enhanced service revenue, its incentive to raise price is strengthened. This "flow-through" effect is shown in Equation 5 by the ratio $(Q_X/Q_{R,X})$ $(\delta P_X/\delta P_R)$. The denominator is the increase in exchange carrier's internal costs due to an increase in the equilibrium price of the basic service, and the numerator is the resulting increase in enhanced services revenue stemming from the change in the equilibrium price of the enhanced service. It is very unlikely that this ratio can ever equal 1. As was discussed earlier, marginal VARs will exit the industry because their internal cost curves will have increased. However, the increase in enhanced service price will not match the shift in cost curves because the demand for the enhanced service is not perfectly inelastic. Moreover, some marginal VARs may turn to alternative suppliers of infrastructure, thus sustaining the level of production in the industry and restraining the rate of price increase. These reactions mean that the flow-through effect will not equal 1.

Another aspect of Equation 5 is that the more stringent the RoR regulatory constraint is, the lower the feasible price markup. The higher the algebraic value of the RoR regulatory multiplier, λ_1, the tighter the constraint. RoR regulation also has another important but implicit effect on the pricing solution. Because this form of regulation can lead to suboptimal investment in network capital and quality (see discussion later), the computation of marginal production cost itself may be skewed, even though no regulatory multiplier shows up explicitly. The reason is that the cost function from which marginal cost is derived, $C^R(Q_R, K, I_N)$, depends on capital and the rate of investment as well as output. Because marginal cost is the derivative of this function with respect to output, conditional on the prevailing stock of capital and rate of investment, any distortions in these latter variables will contaminate marginal cost estimates.

The effect of price cap regulation on the price markup is shown in the third line of Equation 5. Note that the impact of price cap regulation depends on the size of its regulatory multiplier as well as the RoR multiplier, or $\lambda_2/(1 - \lambda_1)$. What this term indicates is that in a hybrid system with both price cap and RoR regulation, price cap effects are going to be magnified because the price cap multiplier is being divided by a number less than 1.

A key term in line three, Equation 5, is $(\delta\psi/\delta Q_R) - (\delta P_R/\delta Q_R)$. The first term shows how sensitive the price cap is to changes in the level of basic service production, that is, the slope of the price cap schedule. The second term is the slope of the market demand curve for the basic service. These terms are evaluated at the equilibrium pricing solution. If the latter effect is greater in absolute value than the former, then a stable pricing solution will result. For example, an increase in output above the solution point would imply a fall in the market clearing price below the cap. The higher price cap would choke off incremental demand and send the system back to the solution point. If, however, the price cap is more sensitive to the level of production than is the market clearing price, then an unstable solution will result.

Consider a traditional market equilibrium with a downward sloping demand curve. The price cap is assumed to decline with increases in per period production, reflecting productivity adjustments. If the price cap schedule is less steep than the demand curve and is binding at a price below the exchange carrier's monopoly price, then the solution is stable. On the other hand, if the price cap schedule is steeper, then reductions in output will lead the firm to raise its price as it moves back up its demand curve toward the monopoly solution. Conversely, increases in demand would lead to a reduction in the price cap, which would lead to further increases in demand, stimulating a downward spiral of price adjustments and increasing demand. Therefore, a price cap adjustment formula that is overly sensitive to changes in the level of output may produce an unstable solution.

III.B. Investment in Physical Capital

The first-order condition that determines the investment in physical capital, I_N, at each point along the firm's trajectory is:

$$
\begin{aligned}
-(\delta C^R/\delta K) = (U_K + U_C) - \\
[\lambda_1/(1 - \lambda_1)][S\alpha_K(P_K/B) - U_K] + \\
[\lambda_2/(1 - \lambda_1)][(1/B)(\delta\psi/\delta I_N)(r + w - dLn\psi_N/dt + dLnB/dt) + \\
(\delta\psi/\delta K)]
\end{aligned} \tag{6}
$$

Line 1 represents the first-order conditions for the investment decision in the absence of regulation. The left-hand side of line 1 (LHS) is the benefit of adding a unit of effective capital in terms of the perpetual flow of variable resource savings. The right-hand side (RHS) are the user costs of capital associated with both external acquisition and internal adjustment costs. The components of these cost terms are detailed in Table 3.6. They include items such as foregone interest income on the dollars tied up in incremental investment and depreciation charges. Also included are expected rates of change in the real price of capital goods and the real cost of internal resources consumed in the adjustment process, where *real* means quality adjusted. Expectations are included in these cost calculations because current decisions are made partially on the basis of expected price and quality changes. For example, inflationary expectations make it cheaper to undertake investment today. On the other hand, expectations of significant technological advances make it more expensive to invest today in lower quality assets. In the absence of regulation, optimal investment would balance both sides of line 1 at the margin.

Line 2 depicts the effect of rate-of-return regulation on investment. The main point here is that if the effective authorized return on a dollar's worth of (real) capital is greater than the acquisition cost, the firm has an incentive to overinvest. This result is similar to the familiar Averch–Johnson model, although I make a number of additions here. First, the authorized return, S, is weighted by the fraction of capital included in the regulated rate base. If this fraction is set low enough, an incentive to underinvest would be created. Second, the expected rate of change in the quality of network capital is one of the elements affecting the cost side of the investment decision (implicitly included in the cost term, U_K). Third, there will be implicit distortions in the level of production of the basic service and/or the quality variable, B, induced by RoR regulation. Therefore, there will be implicit distortions in the calculation of both investment benefits and investment costs.

Line 3 of Equation 6 shows the effects of price cap regulation on the investment decision. Note that the size of the price cap multiplier is magnified in a hybrid system with rate-of-return regulation, $\lambda_2/(1 - \lambda_1)$, because it is divided by a number less than 1. Thus, hybrid systems exacerbate regulatory inefficiencies.

The first term after the multiplier expression in line 3 is $(1/B)(\delta\psi/\delta I_N)$, the marginal change in the (quality adjusted) price cap triggered by an increase in the firm's rate of investment (or, the marginal investment price cap adjustment, MIPCA). Its sign is positive. Inclusion of this term in the price cap formula by regulators acknowledges that there are cash flow consequences incurred when an exchange carrier undertakes network modernization. Further, the equation shows that the benefit is converted into an annuity flow by $(r + w - dLn\psi_N/dt + dLnB/dt)$. This is nothing more than saying that the price cap adjustment (MIPCA) provides a partial offset to the firm's cost of acquiring capital. Because that cost includes a flow of interest and depreciation charges, among other things, the offsetting benefit is a flow of reduced interest and depreciation charges.

Finally, $dLn\psi_N/dt$ represents the expected rate of change in MIPCA. If the expectation is that regulatory treatment of investment expenditure will improve in the future, such investment will tend to be deferred, for the same reason that expected declines in the real price of capital will defer current investment. Conversely, an expectation of deteriorating regulatory treatment would stimulate current investment. Admittedly, this type of regulatory structure is conjectural, but emanently reasonable.

The final term in line 3, Equation 6 is $(\delta\psi/\delta K)$, the change in the price cap constraint due to an increase in the cumulative stock of network capital. It represents the decline in the price cap associated with an increase in the stock of real capital that enhances the firm's productivity. Thus, it works against the first term (MIPCA). The sign of this term is assumed to be negative.

The net effect of price cap regulation on the investment incentive therefore depends on the relative sizes of the marginal investment price cap adjustment

(MIPCA) and the marginal capital stock adjustment, as well as expectations of future regulatory treatment, that is, changes in MIPCA.

III.C. Investment in Innovation to Improve Network Quality

The following equation depicts the firm's innovation decision or choice of network quality at each point along its trajectory:

$$-(\delta C^R/\delta I_N + P_K)(\delta I_N/\delta B) = U_B +$$
$$[\lambda_1/(1 - \lambda_1)][P_K(\delta I_N/\delta B) - S\alpha_K P_K(\delta K_N/\delta B)] +$$
$$[\lambda_2/(1 - \lambda_1)][(\delta \psi/\delta I_B)(r + h - dLn\psi_B/dt) + (\delta \psi/\delta B) + (\delta \psi/\delta I_N)(\delta I_N/\delta B)] \quad (7)$$

Equation 7 determines the firm's rate of change in the efficiency parameter at each point in time. The parameter B is a one-dimensional representation of the quality of network capital and is related to the level of investment in improving network quality (I_B) by the law of motion shown in Table 3.5. I focus on the level of network quality B rather than on the level of investment to improve B because it more closely mirrors how I believe exchange carriers think about these types of investments. Quality can refer to the effective traffic capacity of a physical asset, route diversity, and redundancy, as well as embodied software capabilities for maintenance and network administrative functions.

Line 1 of Equation 7 shows the first-order condition governing the innovation decision in the absence of regulation. The LHS is the benefit on the last incremental increase in B, in terms of lowering the required amount of nominal investment ($\delta I_N/\delta B$). Reduced nominal investment brings with it corresponding reductions in external capital acquisition costs and internal adjustment costs. The RHS of line 1 represents the user cost of undertaking investment in innovation, and it includes foregone interest income, depreciation of the stock of technical knowledge, and expectations about the cost of resources consumed in the internal adjustment process. Without regulation, innovational activity would balance both sides of line 1 at the margin.

Line 2 shows the impact of rate-of-return regulation on the innovation decision. The expression $[P_K(\delta I_N/\delta B) - S\alpha_K P_K(\delta K_N/\delta B)]$ is the difference between improved cash flow caused by the innovation and the reduction in the firm's allowable gross profit due to the shrinkage of the nominal rate base. In the presence of RoR regulation, innovation has this kind of double-edged effect. Less nominal investment is needed to get the same effective capacity, so cash flow improves, but at the same time the regulated rate base is reduced. The net impact of these two effects essentially is the net profit effect on the firm. This effect exists only because of the mechanics of rate-of-return regulation.

Line 3 shows the impact of price cap regulation on the quality decision. Price cap effects are magnified when there is a hybrid system of RoR and price cap regulation, through the term $\lambda_2/(1 - \lambda_1)$. The first term to the right of the multiplier variables in line 3 is ($\delta \psi/\delta I_B$), the marginal quality price cap adjustment (MQPCA). It is the adjustment in the cap triggered when the firm increases its rate of gross investment in quality, analogous to the investment decision discussed earlier (MIPCA). Here, too, the marginal cap adjustment is interpreted as partially compensating the firm for the increased expenditure made to improve network quality. Because it is an offset to the firm's own internal marginal quality adjustment costs, MQPCA is converted into an annuity flow of savings by the expression $(r + h - dLn\psi_B/dt)$. MQPCA provides partial offsets to foregone interest income (r) and obsolescence costs (h) related to innovational activity. The last term in the parenthesis, $dLn\psi_B/dt$, is very important: It is the expected rate of change in MQPCA. If the firm expects that regulatory treatment for undertaking quality improving innovation is going to be more favorable in the future, such activity will tend to be deferred. Conversely, if regulatory terms are expected to worsen in the future, then current innovational activity will be stimulated.

The next term in line 3, $(\delta\psi/\delta B)$, is the negative price cap adjustment associated with an increase in the cumulative level of B. Because a higher level of B would, loosely speaking, tend to raise the firm's productivity, the associated marginal price cap adjustment would be negative. Finally, the last term in line 3 has two components: $(\delta\psi/\delta I_N)(\delta I_N/\delta B)$. The first shows the increase in the cap triggered by an increase in the rate of investment, discussed in Equation 6, and the second piece shows the decrease in nominal investment needed by virtue of the higher quality of network capital. Hence, as the cumulative level of B increases over time it creates direct downward pressure on the price cap by raising productivity and indirect downward pressure by lowering the amount of nominal investment expenditure needed.

The net impact of price caps on the rate of investment in quality will depend on the interactions between stimulating, flow effects in the price cap adjustment (MQPCA) and the restraining influence of the cumulative value of B. Expectations with respect to future price cap adjustments in MQPCA will also play an important role.

III.D. Joint Decision Making and Pseudo-Efficient Regulatory Systems

The preceding set of first-order conditions were discussed in sequence for expository reasons. However, they are determined jointly and are clearly interrelated. The joint nature of the solution process allows some interesting policy questions to be addressed. For example, suppose price cap regulation did not exist $(\lambda_2 = 0)$. Would it then be possible to modify the parameters of the remaining RoR system such that investment in physical capital and the quality of capital could be optimized at the same time? If so, this would provide policymakers with a way to regulate and at the same time achieve efficient investment and quality decisions—a desirable state of affairs.

To see whether construction of such a pseudo-efficient RoR system is possible, first rewrite the relevant distortions:

Investment distortion (line 2, Equation 6):
$$[\lambda_1/(1 - \lambda_1)] [S\alpha_K(P_K/B) - U_K]$$
Innovation distortion (line 2, Equation 7):
$$[\lambda_1/(1 - \lambda_1)][P_K(\delta I_N/\delta B) - S\alpha_K P_K(\delta K_N/\delta B)] \tag{8}$$

From the policy perspective, rate-of-return regulation would achieve an efficient trajectory for both investment and innovation if the two parts of Equation 8 were to become zero at the same time.

After performing some algebraic manipulations, it turns out that the necessary and sufficient condition for the rate-of-return system in this model to achieve an efficient trajectory is as follows:

$$dLnP_K/dt + dLnK_N/dt = r \tag{9}$$

with the fraction of the capital stock allocated to the regulated rate base:

$$\alpha_K = \{U_K/[S(P_K/B)]\}$$

Equation 9 has a very straightforward meaning. It says that the dollar value of the firm's nominal capital stock expands at a rate equal to the interest rate. If this "golden rule" of capital accumulation is followed, then the rate of return system as described in this model can achieve efficiency for investment and innovational decisions. An important part of this solution is that the fraction of the capital stock allocated to the rate base is equal to the ratio of the user cost of acquiring capital over the authorized rate of return.

How feasible is this course of action? Aside from the considerable practical difficulties of quantifying the components of Equation 8 at each point in time, there is yet a more fundamental problem: The "golden rule" solution in Equation 9 may be inherently unstable. In other words, if the dollar value of the nominal capital stock were to grow at a rate faster than the critical rate, r, the incremental cash flow benefits of undertaking quality enhancements would be more significant to the firm than the contracting influence on the rate base. An incentive to overinvest in quality enhancements would be created, rapidly lowering the quality adjusted price of capital (P_K/B) that stimulates even more investment. Of course, as the growth rate of B increased, the growth rate of nominal capital would slow, but it is not clear that the equality in Equation 9 would be reestablished. Alternatively, if the dollar value of the capital stock were to grow at a rate slower than the critical rate, r, cash flow savings on (a lower amount of) incremental investment would become less important to the firm than the contracting influence on the regulated rate base. Too little quality enhancement would be undertaken, rapidly raising the quality adjusted price of capital (P_K/B), leading to a further curtailment of investment. It is not clear whether the growth rate of B would slow enough to reestablish the (higher) equilibrium growth rate of nominal capital needed for the "golden rule" in Equation 9.

As has been demonstrated, the rate-of-return model is undesirable for a number of reasons related to distortions in input decision making. Nevertheless, the preceding discussions also have indicated the potential for price-cap-induced distortions in investment and innovation as well, specifically:

Investment distortion (line 3, Equation 6):
$[\lambda_2/(1 - \lambda_1)][(1/B)\delta\psi/\delta I_N)(r + w - dLn\psi_N/dt + dLnB/dt) + (\delta\psi/\delta K)]$
Innovation distortion (line 3, Equation 7):
$[\lambda_2/(1 - \lambda_1)][(\delta\psi/\delta I_B)(r + h - dLn\psi_B/dt) + (\delta\psi/\delta B) + (\delta\psi/\delta_N)(\delta I_N/\delta B)]$ (10)

A policy question that immediately presents itself is under what conditions would the distorting effects of price cap regulation be removed. Two conditions are relevant. First, the investment distortion (line 3, Equation 6) totally disappears if:

$$dLn\psi_N/dt + dLn(K_N)/dt = r \qquad (11)$$

Equation 11 has a straightforward interpretation. A shadow value of capital is being defined as $\psi_N K_N$, in which capital is multiplied by the marginal investment price cap adjustment (MIPCA), not the marginal dollar cost. Equation 11 says that this shadow value of capital must grow at the interest rate to eliminate the investment distortion. In fact, the first term of Equation 11 is the expected change in MIPCA. Thus, there is a critical expectations rate. If the expectation regarding future marginal investment price cap adjustments is above the critical rate implied by Equation 11, then today's investment will be curtailed below the optimal amount. If the expectation regarding future MIPCA is below the critical rate, then today's investment will exceed the optimal amount.

The second condition relates to the innovation incentive or the degree of quality (along its various dimensions) incorporated into the network:

$$dLn\psi_B/dt + dLnB/dt = r \qquad (12)$$

Here, a shadow value of the level of quality is being defined as $\psi_B B$. The quality variable is being multiplied by the marginal quality price cap adjustment, MQPCA. Equation 12 states that the shadow value of network quality must grow at the interest rate in order to eliminate distortions in the innovation decision (line 3, Equation 7). The first term in Equation 12 is the expected rate of change in the marginal quality price cap adjustment, MQPCA. If the expectation regarding future marginal quality price cap adjustments is above the critical rate implied by Equation 12, then today's rate of quality enhancement will be curtailed below the optimal amount. If the expectation regarding future marginal quality price cap adjustments is below the critical rate, then today's rate of quality enhancement will exceed the optimal amount.[6]

In the price cap framework presented, the aforementioned stimulation of current investment and innovational activity necessarily will induce higher near-term prices than otherwise because the upward marginal price cap adjustments (MIPCA and MQPCA) would be designed to compensate the firm for the cash flow consequences of its current decisions. Clearly, there is a trade-off between upward pressure on service rates that contribute to the financing of network modernization and longer term subscriber benefits in terms of a more efficient and feature-rich network. Examination of the function presented in Table 6.7B representing the maximum allowable price cap, $\psi(Q_R, K, I_N, B, I_B)$, indicates that as the system stabilizes, the rates of capital investment and innovational activity diminish relative to accumulated stocks. Therefore, the components of price cap adjustments that depend on the cumulative stock of capital and the attained level of quality will take on greater relative importance. In the longer term, at a minimum, increases in the price cap will taper off, and possibly there will be a tendency for downward price cap adjustments to occur.

IV. POLICY INNOVATIONS

Having discussed the distortions that regulation introduces into the first-order necessary conditions for an efficient network evolution path, I now offer a series of conclusions, recommendations, and policy suggestions.

First, to increase the likelihood of achieving a stable pricing solution, the maximum allowed price cap applied to the basic service should be insensitive to changes in the level of production of the basic service.

Second, policymakers must strive to avoid creating strong expectations regarding future adjustments in price cap mechanisms that are tied to investment in capital and quality enhancements. Near-term incentives to invest and improve the quality of network capital will be lowered if there are sufficiently strong expectations regarding the magnitude of future price cap adjustments. It follows that rules compensating the firm for the extra costs of undertaking investment and innovation related activities should be put in place as soon as possible. In practice, attainment of the critical expectations values regarding investment- or innovation-induced price cap adjustments is problematic. Yet, if there is a policy objective of fostering growth and modernization in the telecommunications network, efforts should be expended to develop "exogenous cost factors" in the price cap model that offer partial compensation for investment- and innovation-related expenditure so as to stimulate current activity in the proper direction. Moreover, such an approach would serve to minimize expectations regarding future improvements in regulatory treatment, curtailing the incentive to defer investment and innovational activity. When the externality effects and wide-ranging social benefits of a modernized network are taken into account, it would seem that this kind of purposeful, stimulatory regulatory approach is sound policy.

Third, when one considers the explicit and implicit distortions associated with rate-of-return regulation, the magnification of price cap distortions resulting when a hybrid RoR/price cap system is in place, and the infeasibility and possible instability of a rate-of-return-based efficient trajectory, it becomes clear that stand-alone RoR or hybrid price cap/RoR models embody undesirable policies. In addition to economic inefficiencies, regulatory solutions encompassing RoR and cost allocation rules also impose administrative costs on society. A pure price cap model that, in theory, eliminated jurisdictional cost allocation processes and the patchwork of federal and state systems clearly would save administrative costs, in addition to fostering a more efficient path for network evolution.

Fourth, if the price cap model discussed in this chapter were implemented, and network capital and innovation converged to their "golden rule" accumulation paths, the cumulative capital stock and quality effects on productivity would tend to dominate the mechanics of the price cap formula. The price cap would then tend to stabilize and possibly decrease.[7]

V. SUMMARY

This chapter has presented an analysis of the effects of alternative regulatory forms on an exchange carrier's incentives regarding pricing, investment, and innovational (quality enhancing) activity. The main findings indicate that there are more drawbacks to relying on a rate-of-return or hybrid price cap/rate-of-return model than a pure price cap model in terms of not eliciting the kinds of decisions needed to achieve an efficient evolution of the telecommunications network. Administratively as well as economically it is logical to dispense with the current patchwork system of federal and state regulatory systems and replace it with an integrated, internally consistent price cap framework.[8] Loosely speaking, the stylized version of the exchange carrier model used here might also be viewed as the exchange carrier industry as a whole and the franchised serving area as covering the entire nation. In making this generalization, I am glossing over a number of technical issues concerning consistent aggregation from micro to macro functions. Nevertheless, the general point remains valid. The savings in administrative overhead and the elimination of distortions associated with rate-of-return regulation and cost allocation rules would produce net benefits. The political difficulties of implementing such a system, however, are quite formidable.

ACKNOWLEDGMENTS

This chapter is based, in part, on work I completed for the Economic Analysis Subcommittee of USTA (Broadband ISDN Workgroup) and before coming to Bellcore in February 1990. The views expressed are mine and not intended to represent the views of Bellcore nor any of its owner/client companies. I am grateful to Gerald R. Faulhaber for his thorough review and critique of this chapter. William Lehr also provided many valuable comments and suggestions. I am responsible for any errors.

ENDNOTES

1. The term *basic service* generally implies narrowband. However, in principle it may include a minimal amount of broadband capability, such as access to a gateway.
2. Open Network Architecture refers to an FCC policy initiative designed to make the components of an exchange carrier's network available on a nondiscriminatory basis to all vendors of value-added services. The network components that will be unbundled are still being determined but will include various network access components, switching services, signaling and software services, and so on.
3. Because I will be examining the FONC at each point in time along the optimal path, I will drop the t subscripts from the Lagrange multipliers.
4. In a previous paper I allowed for the allocation of all cost categories between regulated and unregulated sectors and permitted allocation percentages to depend on the mix of services. See Stolleman, "Dynamic Effects of Cost Allocations Between Regulated and Non-Regulated Exchange Carrier Operations," *Proceedings of the Bellcore-Bell Canada Conference on Telecommunications Costing*, San Diego, CA, 1989. These more complex cost allocation rules introduced additional distortions into the firm's decision process by creating incentives to skew the service mix. That analysis also concluded that variable and innovation related costs should be assigned entirely to regulated operations, so that the firm would "see" the true cost consequences of its pricing and investment decisions on the regulated side of its business.
5. Time subscripts on variables such as $Q_{R,t}$ are omitted, and only Q_R is shown to simplify notation.
6. Certain technical, simplifying assumptions were made in order to derive Equations (11) and (12). These relate to symmetry assumptions among the elements in line 3 of Equations (6) and (7).
7. The technically simplifying assumptions in endnote 6 result in a stable price cap when network capital and its average level of productivity (B) both grow according to the "golden rule" in a steady state equilibrium, and there are no expected charges in the parameters of the price cap formula (i.e., MIPCA or MQPCA). In principle, regulators should be able to design the price cap formula so as to produce declining price caps in the future; however, care must be taken to provide enough incentive for the investment necessary to sustain a long-run, stable growth path.
8. See Stolleman, "Policy Position: Alternative Regulatory Frameworks," unpublished manuscript, 1989, on file with the *George Mason University Law Review*.

Chapter 4

Issues in the Pricing of Broadband Telecommunications Services

Bhaskar Chakravorti
William W. Sharkey
Padmanabhan Srinagesh
Bellcore

1. INTRODUCTION

In this chapter, we consider some of the long-term issues associated with the pricing of broadband telecommunications services. Future broadband digital networks will be based on a technological platform that will support voice, data, image, and video services. Policymakers and regulators have not yet reached a consensus on how these services will be deployed and priced in a competitive market. This chapter seeks to provide economic inputs that may be useful in developing appropriate technological and regulatory policies as these new services are deployed.

Consider first a very simple arithmetical exercise conducted by Robert Pepper of the FCC.[1] A naive, cost-based price structure would price every ATM cell alike, regardless of the traffic it represented. If each asynchronous transfer mode (ATM) cell is priced identically, and this price is chosen so that a local call costs a penny per minute, he argued that a 2-hour video movie (at 45 Mbps) would cost about $843.75. With compression and transmission at T1 speeds, the price for a movie would fall to about $30. This is unacceptably high when compared to substitutes such as videocassette rentals.

Pepper's solution was to suggest that every residential customer should be given an access line with sufficient bandwidth for a voice plus TV channel and be charged a flat rate equal to today's average expenditure on local calls and basic cable service (about $40). This pricing approach opens up some very lucrative arbitrage opportunities. An apartment building could install a PBX (private branch exchange) and order a few access lines, each with the capacity of 672 voice circuits (45Mb/64Kb). These voice services could be resold to tenants for considerably less than $10 per month per tenant and generate huge profits for the reseller. Tenants could purchase video services from other vendors such as cable companies or use antennas to receive over-the-air broadcasts. This arrangement would be considerably less expensive than integrated access and would allow those who do not want cable television to benefit the most.

Arbitrage opportunities such as the one described here could arise if services with very different bandwidth requirements are served by the same technology and are therefore priced similarly. The history of telecommunications shows clearly that when arbitrage opportunities are made available, the market responds. WATS (wide area telephone service) resale, aggregation and resale of Multi-Location Calling Plans (MLCPs), and International Discount Telecommunications' "callback" service all arose from arbitrage incentives built into existing pricing structures. In a broadband context, large customers could use T1 service to send aggregated voice traffic from one location to another. T1 service provides bandwidth of 1.5 Mbs, which is sufficient to carry 24 simultaneous voice calls. Coincidentally, 1.5 Mbs is sufficient to carry compressed video transmissions of acceptable quality. This may suggest that a T1 line can provide an adequate transport mechanism to support video to the home. However, the cost of a T1 line (typically in the range of $500–$1,000 per month) makes it uneconomical for use in providing residential video service. In addition, T1 service allows for two-way transmission and channelizes the available bandwidth so that the 24 calls in progress can be separated at the receiving end. Neither of these functions is necessary for a residential video service. For both economic and technical reasons, T1 is not a platform that simultaneously supports the needs of aggregated voice and video service.

An alternative approach would be to provide a service that is designed to meet the specific economic and technical requirements of residential video entertainment. Asymmetric Digital Subscriber Loop (ADSL) may be such a solution. It provides one-way 1.5 Mbs transport into the home, is not channelized, and is based on different hardware than T1 service, allowing it to be tariffed at a rate that residences may be willing to pay. Furthermore, ADSL-based video service will not meet the technical requirements of aggregated voice transport. This suggests that a product line of services (T1 and ADSL in the prior example) that are designed to serve a spectrum of customer needs can be more effective than a single service.

There is no single "economically correct" model that can be used to price broadband services or to resolve the issues discussed earlier. In the following sections of this chapter we survey a significant body of relevant work that could be used to understand issues related to the pricing of broadband services. We consider the application of pricing methodologies in both partially regulated and in fully competitive markets, and focus attention to the issue of customer resale and arbitrage in light of these economic models. Some of the specific pricing methodologies are presented in mathematical terms. We attempted, however, to convey the relevant ideas in nontechnical language at the beginning of every section. In a brief concluding section we indicate how the results in all of the sections can assist in developing a practical tariffing framework.

2. DEMAND-BASED PRICING

We first consider the standard economic approach to pricing in multiple product firms assuming profit maximization as an objective.[2] If a firm produces a single product, the determination of a profit maximizing price requires a knowledge of the demand and cost functions facing the firm. The demand function is simply a schedule of output quantities that the firm expects to sell at each conceivable price, and the cost function describes the total cost associated with each output level. At a sufficiently high price demand will be negligible so that total profits will be small (even though the profit per unit sold is large). As price is lowered, more units can be sold, and as long as the increased revenue exceeds the increased cost, profits to the firm will increase. The optimum profit maximizing output is the one in which the incremental (or marginal) revenue from an increase in sales exactly matches the marginal cost of an increase in output.

When a firm produces more than one output, similar principles apply, but the firm must now take into account the interactions on both the demand and cost side of increases in any one of its outputs. To describe the profit maximizing rule in this case it is necessary to introduce some mathematical notation. Suppose that $q = (q_1, \ldots, q_n)$ represents a vector of possible outputs for the firm and let $C(q)$ represent the total cost of producing the output vector q. If demands for each of the firm's products are independent, it is possible to write the inverse demand function $p_i = P_i(q_i)$, which expresses the amount that customers are willing to pay for the last unit produced when output is q_i.[3] If the firm is unregulated, profit maximization is achieved by equating marginal revenue with marginal cost in each market. The expression for marginal revenue is commonly expressed in terms of the elasticity of demand:

$$\eta_i = \frac{p_i \, dq_i}{q_i \, dp_i},$$

so that from the equality of marginal revenue and marginal cost one can derive the expression[4]:

$$p_i(1 + \frac{1}{\eta_i}) = \frac{dC}{dq_i}.$$

Because marginal costs are typically greater than zero, it follows that a monopolist will always choose a price at which marginal revenue is positive, which means that demands will always be elastic (i.e., $\eta < -1$). If the demands are interrelated, the appropriate marginal revenue must be adjusted to reflect the effect of a change in the price of one product on the revenues that may be obtained in all other markets.

The formula for the optimal pricing rule for a multiple product monopolist is a special case of the so-called *Ramsey pricing rule*, which could be applied whether or not the firm is regulated. Where a monopoly firm seeks to maximize its profits without any constraints on its level, a regulated firm may have as its objective the maximization of social surplus[5] subject to a budget constraint that is imposed by the regulatory process. The Ramsey pricing rule in the case of independent demands is given by the formula:

$$\frac{p_i - \dfrac{dC}{dq_i}}{p_i} = -\frac{k}{\eta_i}.$$

The number k is chosen to satisfy the budget constraint, where $k = 1$ corresponds to unconstrained profit maximization, $0 < k < 1$ corresponds to budget constrained pricing when there are increasing returns to scale so that prices in excess of marginal cost are required to recover total costs, and $k < 0$ corresponds to budget constrained pricing under decreasing returns to scale.[6]

The Ramsey pricing rule is generally accepted by economists as an appropriate methodology for pricing of heterogeneous outputs. In the pricing of broadband telecommunications services, however, it may not be appropriate to assume that demand functions are independent. These outputs can be either substitutes or complements for one another, and it is necessary for either a profit or surplus maximizing firm to take account of the relevant cross-elasticities of demand. Although the simple formulae defined earlier no longer apply, the derivation of profit and surplus maximizing prices is well understood theoretically and can be readily implemented given appropriate data. These data include estimates of the appropriate marginal costs and estimates of both own-price and cross-price elasticities of demand.[7] This information may be difficult to obtain, particularly for new services that would be offered on a broadband network.

There are two potential drawbacks to the Ramsey pricing methodology in addition to the informational requirements noted earlier. First, Ramsey prices may be perceived as inherently unfair and therefore politically nonviable in a regulatory environment. This follows because the rule requires that the markup of price above marginal cost should be the greatest in those markets in which the elasticity of demand is the least. From the point of view of overall economic efficiency, this rule makes perfect sense because customers with inelastic demands will curtail their consumption less than would customers in more elastic markets. If such a pricing methodology had been applied to traditional telephone services, access to the network and local usage would have borne a significantly larger share of common costs than interexchange toll. It is unlikely that such an outcome would have been accepted by state and local regulators.

The second potential difficulty with Ramsey pricing is that it does not account for the possibility of competition in one or more of the firm's markets. It may well happen that markets with inelastic demands are also served by active or potential competitors, who could profitably beat the Ramsey price. In a fully deregulated marketplace, the presence of competition does not pose any particular difficulties. In this case, the properly interpreted Ramsey pricing rule would take account of the increased elasticity in markets in which competitive forces were most vigorous and accordingly set prices in these markets close to marginal cost.

3. COST-BASED PRICING

For regulated firms, there are theoretical justifications supporting the use of a Ramsey pricing approach as outlined in the previous section. However, as a practical matter, regulated firms are often expected to set prices on the basis of fully distributed costs. In this section we describe a method of cost-based pricing that is, in some sense, the most reasonable among the various possible methods of cost-based pricing.

Cost-based pricing takes as given the vector q of customer demands. Rather than attempting to find the outputs q that maximize profit, or social surplus, cost-based pricing seeks to determine prices p_i that allocate the total cost $C(q)$ in a fair and consistent manner. In this section we demonstrate one method by which a pricing rule can be derived by means of technical properties, or axioms, that one might impose on the set of all conceivable pricing rules.[8] Although this section contains more mathematical notation than most other sections, the mathematics is included only for a precise statement of results. The reader can obtain a general understanding of the methodology without necessarily following the details of the mathematical derivations.

The cost-based pricing approach might be utilized in a regulatory framework, when regulators must consider whether it is in society's interest to allow telephone companies to deploy broadband networks that are capable of delivering broadband services. Because voice, video, and data services will share a substantial amount of common plant and equipment in a broadband network, an important input into any pricing approach is a sensible procedure for the allocation of such common costs. Currently accepted cost allocation methodologies, however, do not imply that every bit must be priced identically. In this section we briefly describe how one such cost-based pricing methodology is defined in the current economics literature.

Cost allocation methodologies can be defined by enumerating *properties* that a reasonable person might want to impose on the set of all possible pricing rules. These properties include ordinary accounting restrictions that are noncontroversial, as well as properties that seek to ensure that the pricing rule is perceived as fair. One set of properties that has been extensively studied is the following:

Property 1 (Cost Sharing): Revenues should exactly recover total cost. We note that total cost includes a payment to equity holders in the firm, which is required in order to allow them to earn a "fair rate of return" on their investment.

Property 2 (Monotonicity): If an increase in the output of a service unambiguously increases total cost, that service should be assigned a positive price.

Property 3 (Additivity): If it is possible to additively decompose the total cost of producing a set of outputs into two or more component cost functions, then the pricing rule should be additive over the component functions.

Property 4 (Consistency): If two commodities have exactly the same effect on total cost, they should be charged exactly the same price.

Property 5 (Rescaling Invariance): If units of measurement are changed, then prices should be rescaled in the natural way.

It has been demonstrated[9] that these five properties define a unique pricing rule, which has a natural interpretation as an average of marginal costs. The so-called *Aumann–Shapley pricing rule*, which the earlier properties define, is given by the formula:

$$p_i^{AS}(\bar{x}) = \int_0^1 \frac{\delta C}{\delta x_i}(t\,\bar{x})\,dt$$

which represents the price assigned to output i when the aggregate output vector $\bar{x}$ is produced and $C(x)$ represents the cost of producing any output x. Thus one sees that the Aumann–Shapley price for output i is the average of the marginal costs of producing an additional unit of output i, as outputs are expanded along the path from 0 to $\bar{x}$.[10] We note that it is possible to define other axiomatic pricing rules that are related to Aumann–Shapley pricing. For a full discussion of these rules, the reader is referred to the papers by McLean and Sharkey cited in the footnotes.

When applied to broadband telecommunications services, the Aumann–Shapley pricing rule defines prices as a function of traffic characteristics such as the frequency of arrival, duration of the call, and the bandwidth requirement.[11] Because the costs associated with traffic intensities of services offered on a broadband network consist of congestion and delay for other services, these costs can also be reflected in the cost-based pricing approach. Let $q = (q_1, \ldots, q_n)$ represent a vector of n "service classes" (e.g., voice, video, data, etc.). Demands for service arrive at a transmission point consisting of k channels and, for simplicity, we assume that the arrival of a "call" of type i is a Poisson process so that q_i measures the probability that an additional call arrives at any instant of time. Calls contribute to overall system congestion in two ways. First, each type i call has a duration, or size, that is exponentially distributed with mean r_i and variance r_i^2. Second, the cost of providing a service depends on the number of processors that are simultaneously required and the different protocols required in transmission. Thus, arriving calls also contribute to system congestion through the number, d_i, of simultaneous channels that are required for the duration of a type i call.

Several different cost functions can be constructed depending on the queue discipline and the buffer size. Queue discipline refers to the order in which arriving jobs are processed. Buffer size refers to the capacity of the system to hold jobs prior to the commencement of service or during service for store and forward applications. Let k represent the number of channels and let B represent the buffer size. Let $g(k,B)$ represent the cost of building a system with k channels and a buffer of capacity B. Typically, g is an increasing function of k and B. Given k, B, and the values of q_i, r_i, and d_i, let $\beta_i(k,B;q,r,d)$ be the blocking probability for a call of type i. Finally, let $w_i(k,B;q,r,d)$ represent the expected waiting time for a call of type i.

We next consider two models that may be used to define a cost function for a general telecommunications design problem:

Model 1: Let $\beta_1^*, \ldots, \beta_n^*$ and $w_1^*, \ldots, w_n^*$ be "acceptable" blocking probabilities and expected waiting times. In this model, the design problem is to minimize the system cost $g(k,B)$ of constructing a facility such that blocking and waiting costs are within acceptable limits. Solving this optimization defines a cost function C as a function of outputs $q = (q_1, \ldots, q_n)$, where each output i is characterized by its service time r_i and its bandwidth requirement d_i.[12]

Model 2: Let c_i be the value to a type i caller if his call is not blocked. Equivalently, c_i represents the economic loss associated with a blocked call of type i. Let γ_i represent the economic loss associated with a unit of time spent waiting in the queue. In this model the system designer wishes to maximize social surplus, which is equivalent to minimizing the sum of capacity cost plus blocking and waiting costs.[13]

A useful special case of Models 1 and 2 is one in which buffer capacity is equal to zero and server requirements are homogeneous, with $d_i = 1$ for all i. Then the blocking probability is the same for all call types and is given by the Erlang loss formula. The cost functions in Models 1 and 2 are then defined by integer optimization problems that can be solved in principle for any vectors of demand parameters q, r, and d. Naturally there are substantial computational difficulties associated with this approach, and specific pricing rules have, so far, been obtained only for even more specialized situations.[14]

Although it does not appear likely that pricing rules based on cost allocation procedures can fully resolve the Robert Pepper conundrum noted in the introduction, the cost allocation approach clearly indicates that average cost per cell pricing is overly simplistic. This follows because the cost function that appropriately models the cost of providing a variety of services depends on the full array of traffic characteristics that characterize the services. Because an Aumann–Shapley price is an average of marginal costs, the Aumann–Shapley pricing rule depends on a complex, and economically meaningful, set of demand parameters, rather than simply on the number of cells that are transmitted.

Cost-based pricing rules have been criticized by economists on several grounds. In their most elementary form, as presented earlier, these rules do not take any account of customer demand elasticities. Furthermore, despite their axiomatic foundations, cost-based pricing rules are inherently arbitrary from a purely economic perspective. That is, they ignore traditional concepts of economic efficiency that relate marginal benefits, or marginal revenues, to marginal costs of production. In addition, cost-based pricing rules are, by definition, unresponsive to competitive pressures that differ in different markets. Thus, cost-based pricing rules have the potential for inviting entry even in situations in which such entry would increase total industry costs. Finally, pricing rules based on cost allocation procedures do not take any account of the potential for customer arbitrage among services. In the remaining sections of this chapter we consider these issues in greater detail.

4. SUBSIDY FREE AND SUSTAINABLE PRICING

In an environment of free entry, demand-based pricing tends toward charging what the traffic will bear, whereas cost-based pricing leads to rules that are completely unresponsive to customer-demand elasticities. In a partially regulated but partially competitive environment, some degree of flexibility, but something less than full flexibility on the part of the regulated firm, appears to be called for as an alternative to either of the approaches previously discussed. The theories of cross-subsidization and of sustainable pricing seek to establish an appropriate degree of flexibility by identifying permissible bounds on prices for individual outputs and collections of outputs.

Thus, the theory of *subsidy-free pricing* is primarily an application of the techniques of cost-based pricing in situations in which there is competition in one or more of the regulated firm's markets. In traditional telecommunications pricing, the allocation of non-traffic-sensitive costs associated with the local loop has been a persistent issue. For these costs, it is well known (and well documented) in the economics literature that all cost allocations are inherently arbitrary, and that reliance on specific fully distributed cost allocation rules in a partially competitive environment can lead to undesirable outcomes, both for telecommunications consumers and for the regulated firm. Nevertheless, there exist in the literature well-established procedures for identifying bounds on permissible cost allocations such that no group of consumers is disadvantaged by any other group of consumers. We consider these issues in this section.

The fundamental principle of the theory of subsidy-free pricing is that no group of customers should pay more for the outputs that it consumes than it would if served by a specialized firm devoted to its needs alone.[15] If it is assumed that each product of a multiproduct firm is consumed by a distinct group of customers, then subsidy-free pricing requires that no subset of customers pays more than the stand-alone cost of serving them. If S represents any subset of customer classes and q^S represents the outputs associated with S, then the subsidy-free conditions can be written:

$$\sum_{i \in S} p_i q_i \le C(q^S).$$

In addition, the firm must continue to break even (including the return to equity holders) so that $\sum_{i \in N} p_i q_i \ge C(q)$. An equivalent way of defining subsidy-free prices is in terms of the "incremental cost" of serving any subset of consumers. According to this criterion, every group should pay at least the incremental cost of serving it so that:

$$\sum_{i \in S} p_i q_i \le C(q) - C(q^{N-S}).$$

where $C(q^{N-S})$ represents the cost of serving all customers other than S. According to this approach, as long as every subset pays enough to cover its incremental cost, any remaining cost can be assigned arbitrarily to any group of customers without violating the principle of fairness implicit in the subsidy-free constraints.

To consider a very simple example, let the cost function be given by $C(q_1, q_2)$ $= f + c_1 q_1 + c_2 q_2$, where f represents a fixed cost of production, and c_1 and c_2 represent constant marginal costs. Such a cost function is the simplest kind of function in which issues of cost allocation arise. In this case a price vector $p =$ (p_1, p_2) is subsidy free whenever $c_i \le p_i \le c_i + f/q_i$ for each service i. A slightly more complicated but also more realistic example is one in which stand-alone cost functions are given as follows:

$$C(q_1, 0) = f_1 + c_1 q_1$$
$$C(0, q_2) = f_2 + c_2 q_2$$
$$C(q_1, q_2) = f_{12} + c_1 q_1 + c_2 q_2.$$

In this case the subsidy-free constraints require that $c_1 + \dfrac{f_{12} - f_2}{q1} \le P_1 \le c_1 + \dfrac{f_1}{q_1}$, and that a similar constraint holds for p_2.

In a free entry environment, in which all potential entrants have access to the same technology, embodied in the cost function $C(q)$, subsidy-free prices also correspond to "sustainable" prices. Sustainable prices are defined as prices that do not invite entry when the industry is a "natural monopoly" (i.e., total costs are minimized when one firm produces the industry output). This is easily seen by referring to the stand-alone test for cross-subsidization. If the stand-alone test does not hold for a particular subset S of consumers, then it would be possible for an entrant to choose alternative prices $p_i' < p_i$ for each customer i and still make positive profits. Of course, this kind of entry is more likely to occur if entry barriers are extremely low and customers are highly responsive to possible small price differences or conditions that may not apply in telecommunications markets. Nevertheless, the theory of subsidy-free pricing defines a framework for pricing in the presence of competitive pressures that may be a useful consideration as a firm faces the complex issue of pricing broadband services.

5. NONLINEAR PRICING
AND THE ARBITRAGE ISSUE

A nonlinear price structure is one in which a consumer's bill is not proportional to the amount he or she purchases. Billing structures consisting of a fixed monthly fee and a fixed usage charge per unit are nonlinear, as a doubling of the units purchased will not result in a doubling of the bill. The price structures for most telecommunications services are nonlinear.

A brief history of the forces responsible for the widespread use of nonlinear prices provides some insight into the arbitrage possibilities that may arise if broadband services are offered. Nonlinear pricing rules have been adopted in the past as a consequence of a regulated telephone company's need to offer volume discounts to its large users. This need arises from the fact that the costs of networks (or the facilities that comprise them) are largely fixed, and the variable costs associated with providing service on a network that is in place are comparatively small. A customer with a sufficiently high level of use will find a tariff structure such as MTS (Message Toll Service) with usage-sensitive charges more expensive than a dedicated facility. Thus, competitive entry into interexchange telecommunications was initially limited to large firms that formed private networks. It is worth stressing that the alternative available to large users involved large fixed costs and no usage-related costs, and that this alternative was typically available on a point-to-point basis. AT&T initially sought to prevent bypass to private facilities by pricing private-line services attractively. As AT&T private lines were provided out of existing facilities that were installed to meet future demand growth, the additional cost of providing these lines was close to zero. The choice between MTS and private lines resulted in a nonlinear price structure, as large users on private lines paid a smaller price (on average, and for additional calls) than did those on MTS.

Volume discounting of switched services was developed along similar lines. MCI introduced its Execunet tariff with the intention of sharing facilities across medium to large users whose traffic was not concentrated in a few routes. AT&T developed WATS service to appeal to the customers who might find Execunet better than either MTS or private-line services. The widespread availability of WATS-like services led to the first major wave of aggregation and resale. It was relatively easy for WATS resellers to set up operations based on inexpensive PBXs that allowed subscribers to dial in, authenticate themselves, and then dial out on a WATS line that connected them to their called party over the public-switched network. The extent of this resale market was probably unanticipated by AT&T. At its peak, the resale market consisted of more than 1,000 WATS resellers. Many have since gone out of business. A factor that probably played a part in the contraction of this industry was the decision by AT&T and the other long distance companies to flesh out their product line by offering a range of options to medium-sized customers. ProAmerica (later ProWATS) and other new products such as Reach Out America and MCI's Friends and Family have reduced the difference in the unit prices paid by large and medium customers. Nevertheless, these differences persist, and some resellers continue to serve niche markets.

As noted in Briere (1990, p. 219),[16] arbitragers are shifting their focus to profitable opportunities created by Multi-Location Calling Plans (MLCPs). These tariffs are designed to meet the needs of customers with offices in many locations, none of which is large enough to benefit from the volume discounts in tariffs such as AT&T's Megacom or ProWATS. The MLCP allows the firm to enroll all its locations in the plan and compute its discount based on the total volume at all locations. MLCP resellers take advantage of this tariff by aggregating customers into collections with enough aggregate volume to benefit from the volume discounts and by jointly applying for an MLCP account. This business is estimated to amount to more than $1.6 billion per year.

The ability of large users to use an alternative access provider implies that local telephone companies can successfully compete only by developing volume discounts aimed at the very largest users. Moreover, in the presence of competition from resellers, economic theory suggests that in order to compete effectively, a provider must flesh out the product line to reach large, medium, and small users. The use of volume discounts is likely to remain important in the broadband environment. It is likely that large suppliers of video services will face bulk tariffs for switched bandwidth that offers them connectivity to their subscribers. Nonlinear pricing theory implies that unless a range of pricing options suitable for medium-sized users is developed, resellers may have the incentive to enter and compete by reselling services intended for video providers.

Another issue is whether voice, video, data, image, and multimedia traffic will each be tariffed independently, or whether packages of switched transmissions services that support multiple applications will be offered. Price structures that offer separate discounts for voice and data (for example) will appeal to customers who have large volumes for either data or voice or both, but not to those who have the same total volume of use yet moderate volumes of each use. Discounts based on total volume across all uses may induce users with high volume in one use, but relatively lower total use, to seek specialized service from competing providers. Whether discounts should be targeted at specific applications or offered based on total use across all applications is a question for further study.

Another important question concerns the form in which volume discounts are offered. Should volume discounts be offered to customers who presubscribe to the appropriate plan and pay one-time installation charges and high monthly fees, or should incremental discounts be offered automatically to customers whose use exceeds prespecified levels? The former approach places the risk of making the wrong choice of plan on the customers. Although telephone company revenues would appear to be higher under this approach, resellers who purchase bulk service from the telephone company and aggregate users in order to minimize risk of usage variation may offer highly effective competition, thus giving the telephone company an incentive to introduce automatic discounts.

We have argued that the development of nonlinear price structures has been motivated largely by competition for large users, but that pricing methodologies should also take account of the potential entry by resellers. Much of the theory of nonlinear prices considers the efforts of a monopolist to segment his or her market through the use of selective discounts.[17] A recent paper by Mandy considers the sustainability of these prices in a competitive market.[18] The main result of the paper is that nonlinear prices will not be sustainable. Competitive firms will seek to reduce market share among those groups paying a low average price by raising the price to these groups, and they will compete for market share among groups paying a high average price by lowering the price to them. This will unravel the nonlinear price, and all groups will pay the same price in equilibrium.

This result is critically dependent on the assumptions that all firms have the same cost structure, that there are no marketing costs, and that there are no quality differentials across firms. The specific ways in which these assumptions are violated will determine the form of nonlinear pricing that could emerge in competitive equilibrium. A clear understanding of cost differences across firms and the marketing costs associated with reaching consumers with limited information is therefore a topic in need of additional research.

6. PRIORITY PRICING AND INTERRUPTIBLE SERVICE

There is now a large literature on the optimal design of product lines in which any one product of the line can substitute for other members of the line. This literature has significant implications for the pricing of services aimed at different applications, although none of the theories specifically consider broadband services.

Consider the provision of applications such as telephony, electronic mail, video services, remote access to host computers, and distributed processing. These applications have differing requirements for underlying network attributes such as security, bandwidth, lost cells, delay, and delay variation. Thus, voice can tolerate a cell-loss probability on the order of 10^{-4}, whereas interactive compressed video requires that the loss probability be on the order of 10^{-10} for acceptable Quality of Service (QOS). Delay and delay variation criteria for voice and video transmissions are roughly 10 msecs. File transfer can often sustain delays on the order of 10 sec while meeting QOS.[19] Therefore, there is considerable heterogeneity in applications' needs for network attributes.

In addition, different customers have differing willingness to pay for these attributes. In the European context, if the network owner integrates vertically into the provision of applications such as electronic mail and video services (i.e., provide content as well as distribution), and if the technical interfaces presented to the customers do not allow for easy substitution across services, then each service can be priced in accordance with the theory of multiproduct monopoly. This theory has been well studied.[20] QOS for each service can be ensured through the use of appropriate congestion control systems and through related resource reservation schemes. This approach would be sustainable only if the network provider could ensure that a line purchased for a particular application (say video conferencing) is not used instead for another application (say tying together two PBXs). As the price per cell is not constant across applications, this kind of arbitrage may be attractive to some customers. An open question is: Should regulators impose heavy penalties for "misuse" of service offerings on the grounds that arbitrage will not allow for cost recovery through efficient market segmentation? What arguments would support this position?

More difficult choices must be made in the current U.S. context, in which line-of-business requirements may preclude network providers from integrating vertically into all stages of production in the information services industry. It is possible that the network operator may then be limited to providing access and transport services alone. Even though all applications run on the same network, their differing requirements for QOS can be supported by offering a product line of access and transport services, with each product in the line offering different qualities in dimensions such as cell delay, cell loss, and priority. Under this theory, differential prices could then be justified on the basis of differential costs associated with different QOS. An important question is: Can a self-selection scheme be used to segment the market in an economically efficient way? The literature on product-line pricing provides a useful framework for the analysis of this issue. Relevant papers include those by Mussa and Rosen,[21] Srinagesh and Bradburd,[22] and Srinagesh, Bradburd, and Koo.[23]

One theme in these papers is that efficient cost recovery requires that the offered spectrum of qualities be wider than a narrow technical analysis would suggest. Mussa and Rosen showed that this expansion of the product set required that the highest quality application be provided with undistorted quality. All other applications should face quality below that provided in a fully competitive market. Srinagesh and Bradburd showed that there are plausible circumstances in which it is optimal to provide lowest quality applications with undistorted quality and to provide superfluous quality to all higher quality applications. Srinagesh, Bradburd, and Koo developed an alternative model in which it pays the firm to offer undistorted quality to consumers in the middle of the spectrum, with quality degradation of lower qualities and quality enhancement of higher qualities. The factor that determines the characteristics of the optimal product line is the correlation between marginal and total utilities across customers. Primary market research on the distribution of willingness to pay across the potential customer population is therefore a critical input in determining the optimal product line.

Many congestion control strategies currently under discussion, such as call control procedures for the setup of virtual connections and flow control procedures, provide a basis for the implementation of product-line pricing of services.[24] Most preventive congestion schemes for connection-oriented networks are based on the notion of a *traffic descriptor*[25] that captures the (statistical) effect of the call on congestion (or network utilization). Examples of traffic descriptors are peak bandwidth requirements, peak to average bandwidth ratios (or burstiness), and duration of burstiness.[26] The literature on congestion has not yet provided a definitive description of this important variable. Call control schemes typically formulate conditions under which a call with a particular traffic descriptor should be accepted.

The traffic descriptor can also be used as a market segmentation mechanism. In particular, we can conceive of different grades of service as being defined in terms of the treatment by the network of calls with different traffic descriptors. Although the engineering view of congestion control focuses on the issue of fairness in handling calls, the economic view would focus on treating different calls differently, with higher priority given to higher priced calls. This scheme could be the basis for successful (in an economic efficiency sense) product-line pricing if traffic descriptors correlated well with willingness to pay. A more general view would make call connection parameters one element of a broadly defined QOS measure.

Another alternative may be to directly mark high price cells with a priority marker.[27] In this scheme, the issue of priority would not be handled only during call setup, but also during the progress of the call itself. One advantage of this procedure is that it will work for a network that handles both connection-oriented and connectionless traffic. Yet another alternative, suggested by Egan,[28] is to use the signaling system to indicate high or low prices based on network congestion and to allow the customers to modulate offered load in response to the price signal. Egan also suggests the use of interruptible service contracts that block (at least partly) some large users' low-priority traffic during congested periods.

In conclusion, we stress two points. It is important to understand the elements of QOS that matter for customer satisfaction if effective market segmentation strategies are to be implemented. It is also important that switch design (buffer management and call acceptance protocols) be guided by the economics of market segmentation.

7. INCENTIVE PRICING
WITH INCOMPLETE INFORMATION

In general, the price of a product should take into account: (a) the costs of manufacturing and supplying it, (b) the customers' willingness to pay for it, and (c) the market structure of the industry in which the product belongs and the prices and output choice made by competitors and the potential for entry into the industry. The choice of an optimal price depends on the firm's knowledge about the many economic parameters underlying these factors. Typically, this knowledge is incomplete. In the absence of a mechanism to gather or elicit information that enhances the firm's knowledge base, the price structure may be less efficient than prices based on full information. These issues arise in the pricing of broadband services.

To give an idea of the kind of mispricing that can occur because of incompleteness of information, we focus on one of the factors previously listed. We assume that the network provider is fully informed about its own costs and technological parameters and about its competitors; however, it cannot directly verify the willingness to pay by customers. An obvious approach to bridging this gap is for the network provider to conduct market surveys to ascertain these values for the different services to be offered on the broadband platform. The information gathered from such surveys is likely to determine not only the tariffing of the services, but also the level of investment in the new fiber-optics network.

Typically, a customer's willingness to pay for a service is determined by the benefits derived from the service. Consider the following highly simplified scenario. Suppose that a network provider contemplates building a broadband network capable of providing new services to a group of 100 subscribers. Each customer i obtains a private benefit of B_i, which could be expressed in terms of dollars and can be interpreted as customer i's "true" maximum willingness to pay. Suppose that the cost of building the network is $1 million, and that this cost has to be fully recovered by a one-time increase in current rates. Suppose that the sum of the true benefits B_i far exceeds the cost of building the network, $B_1 + \ldots + B_{100} > \$1$ million; hence, the network clearly has positive value. On the other hand, it is reasonable to expect that for any i, $B_i < \$1$ million; hence, no customer will find it worthwhile to finance the network by him- or herself.

Suppose that the network's marketing representative plans to conduct a census of all customers in order to obtain an estimate of willingness to pay for new services offered on the broadband network. Each customer is asked to give an indication of the value that he or she places on the proposed network by choosing a number on a scale from 0 to 10. The cost of the network will be allocated across customers as a function of the numbers that are reported. However, if all customers report the minimum valuation, then the network is not built. If the network is built, then, of course, no one can be denied access to the services it provides, regardless of whether they indicated a willingness to share in the cost of its construction. Thus, if Customer i reports 7, Customer j reports 2, and the sum of the reports made by all customers is 500, then Customer i will be charged $1 million times (7/500) and Customer j will be charged $1 million times (2/500). On the other hand, a Customer k who reports 0 pays nothing.

Given the simple cost allocation scheme outlined, how do we expect the customers to report? From the standpoint of any individual customer, it is never rational to report any number other than 0. This is true regardless of what the other customers may have reported. To understand the rationale for this result, consider the following reasoning on the part of Customer i: "If no one else reports a positive number, then the network is not built, and my payoff (in dollars) is 0. But announcing a positive number, say R_i, would yield a negative payoff of $B_i - $1 million, because I would have to finance all of it. If the network is built (i.e., some other customers do report positive values), I will obtain a payoff of B_i if I report 0 and $B_i - R_i / (R_1 + \ldots + R_{100})$ for any report $R_i > 0$. So in every conceivable instance, it is a "dominant" strategy for me to report $R_i = 0$." Every customer rationalizes a report of 0 in this manner, and the network is not built, even though everybody could have benefited from its presence.

The example here is, of course, an extreme case. However, the main point it makes is applicable in all other situations involving incompleteness of information about the preferences of customers: The pricing or cost allocation rules will not be efficient.

The economics literature has expended a considerable amount of energy in tackling the problems associated with incompleteness of information.[29] This area is broadly referred to as the theory of mechanism design, and it finds applications not only to questions relating to allocating the costs of a project such as a broadband network, but also to the design of flow control algorithms for prioritizing users of the network once it is in place.

Once again, instead of giving a broad survey of the literature on mechanism design, we illustrate how the information gap is bridged by such mechanisms using a simple example. We consider the problem of flow control in a communications network, which involves allocating the usage of the network so that an optimal trade-off is reached between submission of jobs to the network and the congestion that results.

Suppose that the network is to be used by different types of customers ranked from those with the highest priority to those with the lowest priority. The priority levels are private information to the customers, and the network administrator cannot observe them. Also, as is evident, it is too costly to audit the customer to obtain an accurate reading of the appropriate priority level. The differences in priority levels translate to differences in marginal utilities to the customers from being allocated a particular arrival rate onto the network and marginal disutilities from the delays generated due to congestion. We formalize the argument as follows.

Suppose that each customer accesses the network at a rate q_i. The aggregate effect of all the customers attempting to use the network is that it leads to a delay, denoted D. Each customer i is characterized by a utility function that is dependent on q_i and D given by $U_i(q_i, D)$. This function is increasing in the first argument and decreasing in the second for all customers; however, for all values of q_i and D, the absolute values of the partial derivatives $\delta U_i/\delta q_i$ and $\delta U_i/\delta D$ are higher for higher priority customers.

The utilities of the customers obtained after a network is built translates into a commitment to pay for the network before it is built. Hence, the network provider's objective is to maximize the sum of the utilities $U_i(q_i, D)$ over all i. This is expected to maximize the aggregate amount that the network operator can expect to raise up front to build the network.

We can imagine the network provider requesting information on the values of $\delta U_i/\delta q_i$ and $\delta U_i/\delta D$ from customers and then adjusting the access rate for each customer in a way so that $\Sigma_i U_i(q_i, D)$ is maximized. Assuming that the network capacity constraint does not restrict choices, the optimal access rate $q^* = (q_1^*, \ldots, q_n^*)$ is achieved when $\Sigma_i U_i$ can be increased no further. This occurs if:

$$\frac{\delta}{\delta q_i} [U_i(q_i^*, D)] = 0 \text{ } for \text{ } all \text{ } i.$$

Of course, the network operator has no direct knowledge of the customers' true values to determine q^*.

One way to solve this problem is for the network operator to have a series of discussions with the users and to ask them for information on their utility functions. Based on these discussions, the permitted access rate for each user i, q_i, is adjusted. This may involve reducing the rate for some users j and increasing it for i. The network operator effectively becomes a clearinghouse for access rates. It can be shown that such adjustments can be made to follow a set of rules so that eventually the objective $\Sigma_i U_i(q_i, D)$ is maximized and the optimal rate q^* is achieved. Essentially, the users behave as if they were players in a game in which the objective is to maximize utility. The adjustment rules as a function of the series of discussions determine the reallocation of utility. It can be shown that as part of the discussions, the users will reveal information about their utilities in a way so that q^* can be determined.[30]

8. CONCLUDING COMMENTS

As was stated in the introduction, there is no single pricing methodology that can be recommended in all circumstances. In this chapter we attempted to outline the approaches that economic theory suggests to be most relevant in pricing broadband telecommunications services. Four specific methodologies were considered: (a) demand-based or Ramsey pricing (including profit maximization as a special case), (b) cost-based pricing using the Aumann–Shapley pricing rule (or one of its variants), (c) nonlinear and product-line pricing, and (d) priority and interruptible service pricing. These rules are not mutually exclusive. For example, nonlinear pricing can be implemented by differentiating customers in a quality dimension in which priority and interruptibliliy of service are important attributes. Furthermore, priority of service can, in principle, be incorporated into the cost-based pricing methodologies because costs of serving different customer classes will depend on their priority level.

Whatever pricing methodologies are adopted, eventually they must be competitive. Therefore, on a per packet basis, a video packet will likely be heavily discounted relative to a voice packet. Voice signals might then be aggregated and packaged so as to resemble a video transmission, assuming that detection by traffic-distinguishing methods is imperfect. As a result, video channels could be used to carry voice traffic at a price that is significantly lower than that being charged by a local operating company for bulk transport of voice. This is one example in which the possibility of arbitrage or resale may have an impact on future pricing methodologies.

We also noted that a telecommunications pricing structure cannot be made arbitrarily complex, as is the case with the pricing of airline services. Telecommunications customers are not likely to tolerate a tariffing system that requires a translation by a computer or an agent (similar to a travel agent). An implication of the need for "simple" pricing rules is that classical economic pricing rules based on marginal cost may not be practical. The marginal cost of a signal in a telecommunications network is basically the cost imposed on the network due to the additional level of congestion. Congestion, in turn, depends on the rate of arrival of packets that constitute a call, the duration of the call, the number of channels used, and the composite effects these attributes have on an alternative call being blocked, delayed, or rerouted. Because the probabilities of such occurrences are constantly changing, the prices should, from an economic perspective, be changing in response. For practical reasons, the prices are set on a much coarser grid to accommodate the customers' aversion to complexity. Moreover, due to externalities between different services on the same broadband platform, the technological parameters do not fully account for the true economic costs. Such externalities are due to a variety of factors: Different services are substitutes for each other (such as voice versus email), and other services complement each other (for example, voice and information services such as electronic yellow pages). In addition, there are congestion externalities. In sum, it is arguable that practical limitations on the pricing structure for services over a broadband network may lead to suboptimal outcomes for both service providers and their customers.

Although this discussion is preliminary in nature, it suggests a useful starting point for telecommunications companies that must consider numerous other factors in pricing in the early stages of broadband deployment. In addition to focusing on the traditional customer base, a parallel effort could be directed toward securing new revenue sources from nontraditional customers. Although it has been widely noted that the convergence of telecommunications and information processing technologies may have implications for traditional entertainment markets,[31] relatively little attention has been paid to the converse proposition—that revenues from nontraditional sources might benefit telecommunications providers under innovative pricing methodologies. In light of these noted difficulties associated with resale and arbitrage, these new revenue sources may play a central role in future approaches to pricing of telecommunications services.

ENDNOTES

1. Pepper, R., *Through the Looking Glass: Integrated Broadband Networks, Regulatory Policies and Institutional Change*, FCC Office of Plans and Policy Working Paper No. 24, 1988, p. 46.
2. An elementary exposition of the theory of multiproduct monopoly pricing is contained in Stigler, G., *The Theory of Price*, Third Edition, London: Macmillan, 1966, p. 211. A more rigorous treatment may be found in Tirole, J., *The Theory of Industrial Organization*, Cambridge, MA: MIT Press, 1989, p. 69.
3. When demands are not independent, the inverse demand function depends on all other quantities produced by the firm. We note that our analysis also ignores income effects, which is reasonable in an analysis of telecommunications demand functions.
4. Because total revenue in market i is given by $P_i(q_i)q_i$, profit maximization requires chosing output q_i such that:

$$\frac{d}{dq_i}[P_i(q_i)q_i] = \frac{d}{dq_i}[C(q_1, \ldots, q_n)].$$

5. Social surplus is defined as profits plus aggregate consumers' surplus, in which the latter is the difference between the amount that consumers would be willing to pay for a given output, given an all-or-nothing offer, and the amount they are asked to pay. This quantity is computed by integrating the appropriate demand function.
6. More detailed discussions of Ramsey pricing may be found in Sharkey, W. W., *The Theory of Natural Monopoly*, Cambridge, UK: Cambridge University Press, 1982; Brown, S. J. and D. S. Sibley, *The Theory of Public Utility Pricing*, Cambridge, UK: Cambridge University Press, 1986; Tirole, *op. cit.*; and, Mitchell, B. M. and I. Vogelsang, *Telecommunications Pricing: Theory and Practice*, Cambridge, UK: Cambridge University Press, 1991.
7. Cost functions appropriate for broadband telecommunications services are considered in R. P. McLean and W. W. Sharkey, "An Approach to the Pricing of Broadband Telecommunications Services," *Telecommunications Systems*, forthcoming. Demand elasticities may be determined from econometric techniques applied to survey data on anticipated demand or to actual demand for closely related services.
8. This approach, however, does not take account of political, regulatory, or public policy considerations.
9. See Miriam, L. and Y. Tauman (1982), "Demand Compatible, Equirable, Cost Sharing Prices," *Mathematics of Operations Research*, 7, 45–56. A somewhat different axiomatic characterization of Aumann–Shapley pricing is given in Billerica, L. J. and D. C. Heath (1982), "Allocation of

Shared Costs: A Set of Axioms Yielding a Unique Procedure," *Mathematics of Operations Research*, 7, 32–39.

10. A more complete discussion of the Aumann–Shapley pricing rule and its application to some specific cost functions is contained in R. P. McLean and W. W. Sharkey, "Alternate Methods for Cost Allocation in Stochastic Service Systems," unpublished manuscript.

11. Specific functional forms of such pricing rules are contained in the papers by McLean and Sharkey, *op. cit.*

12. Formally, the cost function C is defined by:

$$C(q,r,d) = \operatorname*{Min}_{k,B} g(k,B)$$

subject to

$$\beta_i(k,B;q,r,d) \le \beta_i^* \text{ for all } i$$
$$w_i(k,B;q,r,d) \le w_i^* \text{ for all } i.$$

13. Formally, the cost function $C(q,r,d)$ is given by the following expression:

$$C(q,r,d) = \operatorname*{Min}_{k,B} g(k,B) \left[g(k,B) + \sum_{i \in N} c_i q_i \beta_i(k,B;q,r,d) + \sum_{i \in N} \gamma_i w_i(k,B;q,r,d) \right].$$

14. See papers by McLean and Sharkey, *op. cit.* The Erlang loss formula that defines the blocking probability is given by:

$$\beta(q,r,k) = \frac{(q \cdot r)^k / k!}{\displaystyle\sum_{i=0}^{k} (q \cdot r)^i / i!}.$$

Because the buffer capacity B is constrained to be equal to 0, there are no waiting costs, and the cost functions in Models 1 and 2 become, respectively:

$$C(q,r,d) = \operatorname*{Min}_{k=0,1,2,\dots} g(k) \text{ s.t. } \frac{(q \cdot r)^k / k!}{\displaystyle\sum_{i=0}^{k} (q \cdot r)^i / i!} \le \beta^*$$

and

$$C(q,r,d) = \operatorname*{Min}_{k=0,1,2,\dots} g(k) + (c \cdot q) \frac{(q \cdot r)^k / k!}{\displaystyle\sum_{i=0}^{k} (q \cdot r)^i / i!}.$$

15. This concept clearly is not intended to apply in all situations. Technically, it is appropriate in a technological environment of increasing returns to scale, in which additional consumption by one group of consumers does not necessarily make every other consumer worse off. In a decreasing returns situation (e.g., fishing from a common ocean or drawing water from a common aquifer), different principles must be applied. Even in an increasing returns environment, subsidy-free prices may fail to exist, as has been described in Sharkey, *op. cit.*

16. Briere, Daniel D., *Long Distance Service: A Buyer's Guide*. Norwood, NJ: Artech House, 1990.

17. A comprehensive survey can be found in Wilson, R., *Nonlinear Pricing*, London: Oxford University Press, 1993. The issues of risk are treated in Clay, K., D. S. Sibley, and P. Srinagesh (1992), "Ex Ante and Ex Post Pricing: Optional Calling Plans and Tapered Tariffs," *Journal of Regulatory Economics*, 4, 115–138.

18. Mandy, D. M. (1992). "Nonuniform Bertrand Competition," *Econometrica*, 60, 1293–1330.
19. See Hong, D. and T. Suda (1991, July), "Congestion Control and Prevention in ATM Networks," *IEEE Network Magazine*, p. 11.
20. See, for example, Brown and Sibley, *op. cit.*
21. Mussa, M. and S. Rosen (1978), "Monopoly and Product Quality," *Journal of Economic Theory*, pp. 301–317.
22. Srinagesh, P. and R. Bradford (1989), "Quality Distortion by a Discriminating Monopolist," *American Economic Review*, pp. 96–105.
23. Srinagesh, P., R. Bradford, and H-W. Koo (1992, June), "Bidirectional Distortion in Self-Selection Problems," *Journal of Industrial Economics*, 40, 223–228.
24. See Hong and Suda, *op. cit.* and Vakil, F. and H. Saito, "On Congestion Control in ATM Networks," *IEEE LCS Magazine*, forthcoming.
25. See Vakil and Saito, *op. cit.*
26. See Hong and Suda, *op. cit.*
27. See Hong and Suda, *op. cit.*, pp. 14–15.
28. Egan, B. (1987), "Costing and Pricing the Network of the Future," *Proceedings of the IEEE, ISS*, pp. 483–490.
29. For recent surveys of the field, see Moore, J., "Implementation in Environments with Complete Information," and Palfrey, T., "Implementation in Bayesian Equilibrium: The Multiple Equilibrium Problem in Mechanism Design," both in *Advances in Economic Theory*, ed. by J.-J. Laffont, Cambridge: Cambridge University Press, 1993.
30. For details on how such an adjustment rule works, see Chakravorti, B., "Optimal Flow Control of an M/M/1 Queue with a Balanced Budget," *IEEE Transactions on Automatic Control*, forthcoming. A survey of this literature is contained in Sharkey, W. W., "Network Models in Economics," *Handbook of Operations Research and Management Science: Networks*, forthcoming.
31. See, e.g., Curtis, T. and K. Means, "Market Segmentation and the IBN Policy Debate," in *Integrated Broadband Networks: The Public Policy Issues*, ed. by M. C. J. Elton, Amsterdam: North-Holland, 1991.

Part II

REGULATORY PRACTICE

Chapter 5

A New Index of Telephone Service Quality: Academic and Regulatory Review

Sanford V. Berg
University of Florida

1. INTRODUCTION

The academic ideal, for many scholars, is to do research that is simultaneously theoretically elegant and solves important practical problems in the real world. In this chapter, I provide a case history of my own attempt to pursue this ideal. My work focused on the regulatory measurement and reward of service quality provided by regulated local telephone service companies operating in the state of Florida. I describe the challenges my research team faced in (a) maneuvering a relatively straightforward regulatory innovation into regulatory practice, and (b) disseminating the conceptual framework to another community (through publication in scholarly journals read and edited primarily by academics).

Although others will evaluate the ultimate success of this dual research strategy, there are lessons to be learned from the effort. Generalizations from one observation are problematic—but even one data point can shed light on the issues associated with the adoption of rules and procedures which improve resource allocation in telecommunications. This work suggests that both sound theory and supporting empirical evidence are necessary if current approaches to quality are to be strengthened. I offer some low-brow theory and observations on actual regulatory behavior to draw lessons regarding the role of academic research in promoting regulatory innovations.

2. OVERVIEW OF CURRENT REGULATORY
SCHEMES AND THE PROPOSED INDEX

The importance of service quality has been highlighted by developments in the last decade: divestiture, network interconnection, and technological change (Rovizzi & Thompson, 1992). The competitive and complementary service offerings of new entrants raise challenges for incumbent local exchange carriers. In the meantime, state regulators are still faced with the choice between traditional cost-of-service regulation and various forms of incentive regulation. Whatever their decision, the role of quality has a higher profile than in the past.

The regulatory process utilizes technical performance features of networks, even while recognizing that consumer satisfaction may depend only indirectly on engineering measures of service quality. These surrogate evaluations tend to consist of pass/fail technical standards that were often established decades ago. They have grown by accretion: The most recent National Association of Regulatory Utility Commissioners (NARUC) compendium on the subject lists between 90 and 100 separate standards (depending on how one groups some subcategories).

Current quality-of-service pass/fail targets are somewhat arbitrary, having arisen from a chaotic process reflecting historical engineering capabilities, political pressures, and administrative happenstance. Consumer valuations of different quality dimensions and corporate recognition of emerging technological opportunities are not likely to be captured by pass/fail standards. In addition, combining information on multiple dimensions into an overall assessment is very difficult for regulators.[1] Information overload could lead to "management by exception." By focusing on the rules that a company fails, regulators essentially ignore dimensions on which the company being evaluated has exceeded the standards. Similarly, the use of cutoff targets gives companies currently operating below a standard no incentive to improve performance if those improvements would still leave them short of the standard. Thus, perverse incentives result from the use of pass/fail standards. Developing an appropriately weighted quality-of-service index is no simple task, but the approach represents a potential improvement over multiple pass/fail quality standards.

2.1 Production Possibilities for Pass/Fail Standards

Current reward schemes, in Florida as in other states, compare a company's objective scores, $Z_1 \ldots Z_n$, on a set of engineering attributes to standards, $Z_1^* \ldots Z_n^*$, set by the regulatory agency on those attribute dimensions. Below-standard performance on any attribute triggers censure, whereas performance above any standard is not treated as being any better than performance exactly at the standard. This tacitly drives companies to produce quality along each attribute at exactly Z_i^*, because exceeding Z_i^* generally takes resources but is not rewarded in the regulatory system. Beyond this, however, very little could be said about exactly what regulators were rewarding or trying to reward, as they used subjective and intuitive judgment to assess the overall service quality of a firm that had a complex set of above-standard and below-standard scores on the many measured dimensions of quality.

We attempted to make their expert judgment more systematic by modeling regulators' trade-offs among various dimensions of quality, $\hat{Q} = f[(Z_1 - Z_1^*), \ldots, (Z_n - Z_n^*)]$. We discovered that regulators often agreed that overall quality was higher when Standard A was exceeded and B was failed (when Attribute A was considered relatively more important) than when A and B were met exactly. Therefore, existing evaluation policies were creating perverse incentives by treating the former company as being in violation and the latter as being in compliance with regulations.

In our proposed alternative index, we define $\hat{Q}^*$ to be the predicted level of overall quality associated with meeting all n quality standards exactly. We then propose that any combination of $Z_1 \ldots Z_n$ that leads to $\hat{Q} \geq \hat{Q}^*$ should be treated as meeting the standard. Under this regime, companies would offer whatever combination of substandard and superstandard $Z_1 \ldots Z_n$ that allows them to achieve $\hat{Q}^*$ in the most efficient way given that company's cost structure. For each 1% point above a standard, the quality index rises—depending on the weight given to that particular standard. Similarly, shortfalls result in reductions in the index.

For simplicity, consider two dimensions of service quality monitored by regulators: dial-tone response (Z_1) and call completions (Z_2). Like other commissions, the Florida Public Service Commission (FPSC) has standards for each of these. Florida requires that 95% of all calls receive a dial tone within 3 sec. The intraoffice call completion standard requires the successful completion of 95% of all calls to numbers with the same first three digits as the calling number. A welfare-maximizing regulator will induce the firm to equate the marginal benefits from service quality improvements with the marginal costs. This condition for optimality is depicted in Fig. 5.1 (adapted from Berg & Lynch, 1992). Three production possibility frontiers (PPFs) are shown—those that are further out require that additional resources be devoted to the production of quality: $100, $110, and $130, respectively. For a given level of real resources, improvements in one quality dimension involve a deterioration in the other. The PPF reflects engineering and resource constraints. The slope of the PPF represents the opportunity cost of increasing Z_2: Given the constraints, there must be a reduction in Z_1.

2.2 Relative Valuations of Service Characteristics

Relative valuations for the two dimensions of service quality are also shown in Fig. 5.1. In this example, the subjective trade-offs by customers are reflected in the preference mapping characterized by $U = 2,010$ and $U = 2,020$. That is, for any given level of satisfaction (for example, $U = 2,010$), if one dimension deteriorates (say Z_2—call completions—falls from 95% to 90%), then Z_1 must increase if customers are not to be made worse off. Here, Z_1 (dial-tone response) must rise from 95% to 97% if the customers are to remain on $U = 2,010$.

In this example, points E and M represent the same level of satisfaction, met by different combinations of service qualities. Point E would not be a welfare maximizing point because point M is valued equally by consumers and costs less to achieve. At point E, the subjective marginal rate of substitution between Z_1 and Z_2 does not equal the marginal rate of transformation (as reflected in the slope of the production possibility frontier). The proposed approach would drop the pass/fail standards of 95%, 95%, and give the telephone company flexibility in selecting least-cost ways to achieve a given level of performance. In the example, point X could be achieved for the same resource cost as point E—but benefits would now be $U = 2,020$.

Figure 5.1 illustrates how firms could be presented with a regulatory objective function and allowed to trade off high-cost (low-valued) quality dimensions for low-cost (highly valued) quality dimensions. In the simple example, if (95,95) yielded an "acceptable" overall level of quality, one scoring function that would signal the telco to modify its quality mix would be $Q = Z_2 + (5/2)Z_1$, and the minimum quality "score" is $Q = 332.5$. The firm would go to point M (97, 90).

One issue is whether the minimum quality "score" is appropriate. In the simple example, if $Q = 337.5$, the firm will be driven to point X instead of point M. If point M corresponds to \$105, and an additional \$5 lets the firm attain point X, then the outlay is worth it if X is valued \$5 or more than the quality bundle at M. We did not try to attack the incremental cost/incremental valuation issue, but focused on replacing the (95,95) standard with a minimum quality score of 332.5. The question is how to derive the weights described earlier.[2]

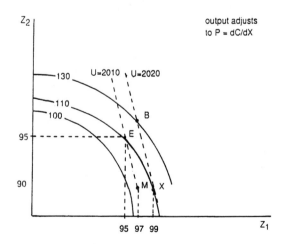

FIG. 5.1. Relative valuations of quality attributes.

This is an extremely simple idea that can be easily shown to represent an improvement over the current regulatory regime, so long as there is agreement as to the weights to be given the various standards (Noam, 1991). Yet, the legal, political, and institutional roadblocks to its full adoption and appropriate implementation have been numerous. At the same time, we have published and presented this work in several academic venues. However, the primary theoretical piece is still battling to emerge from a 4-year review process at a prestigious academic journal that strives to merge theory and practice.[3]

3. SCIENTIFIC REVIEW AND THE CREATION OF KNOWLEDGE

Publication lags, like regulatory lags, arise from the existence of numerous checkpoints in which stringent review criteria are applied. Academic gatekeepers provide critical reviews of analyses, giving readers some confidence that the published article represents a contribution to the literature.[3] External reviewers can help researchers focus their efforts and remedy potential flaws in analyses. The review process screens potential contributions, asking whether the submission contributes new and creative insights regarding the issue at hand. A second and perhaps more fundamental question is whether the issue under consideration is actually important.

Theoretical constructs, empirical tests, and historical evaluation provide the three legs upon which a policy science stands. Given the gains from specialization and division of labor, researchers will tend to tackle issues from one of the three perspectives, although good analysis using any one of the three modes cannot ignore the other two. For example, good theory recognizes the historical setting that establishes the institutional context for theoretical analysis. In addition, theory often depends on empirical observations (in the form of stylized facts) to provide bounds on key parameters or to determine the signs of particular relationships. Reviewers of potential contributions know the economic paradigm from which models are derived and quantitative tests conducted. Deviations from the widely accepted neoclassical economic framework face a hurdle in the review process. Because this work does not draw heavily on the paradigm of the rational, evaluative, maximizing consumer, I have had to justify the framework to skeptical academic reviewers.

The concern here is with the evaluation of telecommunications service quality. The literature on product (or service) quality is voluminous. Because the theory is summarized elsewhere (Berg & Lynch, 1992), it will not be surveyed here. Suffice it to note that the models are elegant, often insightful, and difficult to test. Quality outcomes under competition, monopoly, and regulation depend on a host of factors, including incremental costs, incremental benefits, and average benefits associated with quality changes. Furthermore, the introduction of multiple dimensions of quality greatly complicates the analysis.

Nevertheless, the neoclassical paradigm suggests that informed consumers will evaluate alternative service offerings and select consumption bundles based on their preferences. Thus, the ultimate judge that matters is the rational customer, not some regulatory surrogate. I agree. However, policy analysis cannot abstract from the institutional context: The past matters. Regulatory reviews are designed to screen for a different set of problems than those that might concern scientists. My work has tried to be responsive to both review processes. Indeed, the studies have benefited from discussions with academics and regulators. Nevertheless, it is somewhat risky for academics to adopt a research strategy that tries to meet criteria from both review processes.

4. REGULATORY REVIEW AND PROCEDURAL FAIRNESS

The regulatory review process is grounded in procedural fairness. Disruptive transitions are costly to buyers and sellers alike, so administrative delays can sometimes serve as mechanisms for smoothing out the impacts of unanticipated changes in demand or technologies. Review lags ensure that proposed changes are understood by all those affected by new regulations. In addition, administrative procedures are designed to provide opportunities for complaints to be heard. Hearings and informal workshops serve as forums in which stakeholders present their concerns.

The heavy role of legal, accounting, and engineering expertise at public utility commissions suggests that economics by itself provides an inadequate foundation for regulatory decisions. Emphasis on legal precedent gives continuity to the rate-making process—forcing some consistency in the face of emerging problems. Protection is afforded both the regulated firm and its customers. From the standpoint of service quality, fairness toward customers requires that data be verifiable. If reported data are fabrications or the result of improper manipulation of data collection procedures, the firm loses credibility. Credibility is essential if the regulatory process is to be accepted by consumers.

Similarly, the heavy dependence on accounting data constrains much of the debate to the consideration of actual outlays, rather than hypotheticals. Although future test years are utilized in many jurisdictions, the focus is still on accounting rather than economic costs. Economic opportunity costs may be considered, but accounting data and cost allocation procedures based on historical developments dominate rate cases so long as there are no alternative suppliers. In addition, engineering data are particularly relevant for considering quality-of-service issues because these are objective and subject to review.

The regulatory concern is that regulated (or partially regulated) firms may choose the wrong level and mix of quality, recognizing that quality is multidimensional. But what does *wrong* mean? Too little quality? Too much quality?

An inappropriate mix of quality components? An inappropriate pricing of quality components?

The framework described here is appropriate for encouraging the right mix of quality characteristics. By itself, it does not address the overall level of quality. However, the current battery of pass/fail standards addresses neither the mix nor the level of quality. By building on current data collection efforts, the proposed approach maintains continuity. In addition, the framework allows regulators to formalize what they mean by quality. The procedure for soliciting weights has its limitations. However, the weights can be refined. With the adoption of a quality index, managers can make network investment and operations trade-offs based on precise weights applied to a specified set of characteristics. This represents an improvement in the process because the current pass/fail targets are subject to uneven regulatory application. The current process can be quite inefficient as well as potentially unfair.

Thus, from the standpoint of procedural fairness, the proposed quality index is quite promising. As with any new instrument, however, telecommunications firms will be hesitant to accept new rules. For example, two problems with the movement to price caps are the determination of the starting price level and the calculation of the productivity adjustment. Both would have to be determined in advance before a telco would support the replacement of rate of return on rate-base regulation with price caps. Similarly, before firms will accept a new quality index, they will want to know how it is to be used in the regulatory process. If it is seen as another tool for bludgeoning the firm, the index will not receive their support. Strong opposition by regulated firms reduces the likelihood of adoption.

Intervenors will also be skeptical of any departures from current conventions. For example, the Public Counsel's Office will primarily be representing residential customers. The weights appropriate for these customers may not be the same as those for high-volume commercial and industrial demanders. Because the dimensions of quality tend to be collective consumption goods (quality available to one is available to all), consumer advocates may not want to give the firm the discretion involved in meeting an overall quality-index constraint. Rather, they may prefer to focus on items of particular concern to their constituency.

Given these observations regarding the focus on continuity and aversion to change, stakeholders will delay adoption of new instruments by regulatory agencies until they have made thorough checks on implications for performance. Regulators, utility managers, and consumer advocates will all need to be comfortable with the new approach.

5. CASE STUDY OF A NEW QUALITY INDEX

The rationale behind the proposed quality index, Q, has already been described. Lessons can be learned by reviewing the parallel evolution of the conceptual framework and associated regulatory rules. Let us begin at the beginning: The investigation was initiated in late 1986, when FPSC staff approached the Public Utility Research Center (PURC) about exploring ways to evaluate quality of service provided by local exchange companies. Based at the University of Florida, PURC attempts to bridge principles and practice, so the issue clearly fits into its purview.

5.1 Initial Data Collection

Academic institutions are not consulting groups who can switch resources rapidly from one activity to another. The author tried to identify researchers who could assist with the project. Two marketing professors had ideas about how a more comprehensive indicator of quality might be developed. Interactions with FPSC staff yielded what was thought to be a good understanding of the pass/fail standards applied to telcos. Teaching responsibilities and other research commitments meant that much of the initial investigation occurred in the summer of 1987. Figure 5.2 provides a time line for the stages of regulatory and academic reviews associated with the proposed framework.

The basic methodology involved a survey in which experts made comparisons of quality bundles. The weighting scheme was developed by having experts from the FPSC rate different hypothetical company profiles of performance on the rules within nine rule clusters. Each profile was rated on a scale from 1 (worst possible performance) to 10 (best possible performance). Similar comparisons were made across rules so that a comprehensive score could be assigned to a telco based on its observed performance on the 38 dimensions. The entire procedure is a form of conjoint analysis called the hierarchical conjoint analysis (Louviere, 1984). (See the Appendix for an example of how weights were derived.) This approach is suited to capturing the trade-offs experts make in overall evaluations of objects that can differ on a very large number of attributes that can be logically grouped into subsets of related attributes.

In January 1988, a report was ready for the FPSC (Buzas & Lynch, 1988). The methodology for determining weights to be given the various dimensions was outlined in some detail, and associated statistical tests were presented. It provided illustrative calculations for hierarchical conjoint analysis so the derivation of individual weights could be described. The study analyzed how agreement and disagreement among survey participants could be identified. The report included a discussion of each item on the questionnaire, defined technical terminology, and showed which of the 38 dimensions had greatest weights, based on responses provided by FPSC staff. The formula derived showed the weight of a 1 percentage point change on each of the 38 dimensions.

The three-member research team prepared an academic working paper in July 1988. This paper was the first project output focusing on both the methodological and policy issues associated with the new index. The weights were calculated and applied to a hypothetical telephone company. The implications of agreement

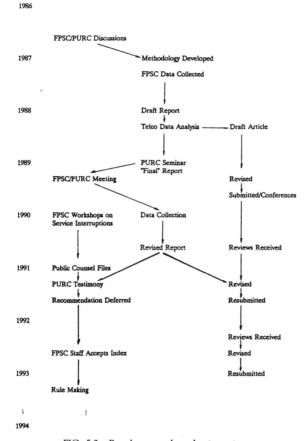

FIG. 5.2. Regulatory and academic reviews.

and disagreement among survey participants were also outlined. The report contained examples of techniques for determining the reliability of estimates.

Finally, limitations to the study were identified. In particular, the separate role of costs was discussed. Also, because the initial trade-offs were made by FPSC employees, it was noted that these experts might have a different perspective than either consumers or companies. Third, the trade-offs were made without reference to specific locations or clientele. It is plausible that the FPSC might want to reward compliance on some rules more heavily or lightly in certain geographic areas. For example, compliance on public telephone dimensions might be more important in rural areas than in urban areas because the phones are further apart in the former.

The research team needed to determine whether experts at telephone companies would give the same weights to the various rules. We thought that telco representatives might be more aware of the relative benefits of meeting the different rules. Alternatively, different firms face different mixes of customers (due to demographics, per capita income, degree of urbanization). Such factors could mean that customer valuations for different components of quality differed across firms, making a single weighting scheme inappropriate. Despite the remaining issues, the team believed that the results were highly suggestive: Progress was being made in this very complicated area. The research team viewed the methodology as offering a way to introduce greater rigor and content into the quality evaluation process. The project turned to the issue of telco weights.

The data collection effort moved forward in earnest, as the team sought cooperation from regulated firms. Some were willing to devote personnel to the effort, but others were concerned about implementation issues. A PURC seminar was held in February 1989. Berg, Buzas, and Lynch described the rationale and methodology behind the comprehensive index. Results from the FPSC sample (12 employees) and two companies were presented. Alan Taylor, Chief of the Bureau of Service Evaluation, represented the FPSC. Also in attendance were representatives from the major Florida telcos. These representatives expressed a concern as to how the proposed index might be used: to evaluate firms at a single point in time? to evaluate trends over time for a single firm? to compare firms? The different service territories and degree of network modernization influence the starting point for each firm, raising a concern for fairness. Cross-firm comparisons might not take into account different technological opportunities. Executives tended to see the index as another factor that could be used against them at the next rate case.

Most of the formal presentations focused on basic methodological points:

1. Different dimensions had different weights.
2. Only a few dimensions really mattered a great deal.
3. There was a very high correlation (Pearson correlation coefficient of better than .90) between weights obtained from the two companies and from the commission.

A hypothetical example was given, and the method of calculation was presented. To illustrate the usefulness of a comprehensive index, data for two hypothetical companies were presented, emphasizing the difficulty of making pass/fail comparisons. The politics of regulation was not given much attention, although attendees were probably more worried about implementation issues than statistical refinements.

The initial scoring function had 38 weights plus a base "score" if each standard as exactly achieved:

$$Q_a = 5.92 + .1172(Z_1 - Z_1^*) + .0786 (Z_2 - Z_2^*)$$
$$+ .0813(Z_3 - Z_3^*) + \ldots + .0198 (Z_{38} - Z_{38}^*)$$

Here $Z_1^* = 95\%$ of calls that received a dial tone in 3 sec

$Z_2^* = 95\%$ of intraoffice (same first three digits) calls completed

$Z_3^* = 95\%$ of interoffice calls completed

•

•

•

$Z_{38}^* = 100\%$ of all public telephones that have their locations posted, and the identifications of locations coordinated with the appropriate 911 or emergency center.

For this example, a telephone company just meeting each standard would score 5.92. A company score of 96% on dial tone delay (Z_1) would more than offset a company score of 94% in interoffice call completions (Z_2), in which the standard is 95% for each but the weight for the former is greater than the weight for the latter. Note that 16 of the 38 weights in this initial scoring function referred to public telephones: functioning (receives calls), enclosures (handicapped access, cleanliness, and lights), coin operations (coin returns, operator assistance), directory availability, and so on. In addition, the linear form of the scoring function implied that a 1 percentage point improvement for a service dimension had the same impact whether the change was from 90% to 91% or from 96% to 97%. This was approximately true regardless of whether the change represented a movement toward the standard or one which exceeded the standard.

Within weeks, the Final Report on Telephone Service Quality was sent to the FPSC. The academics thought their jobs were done. Some project results were reported in a conference volume that appeared 2 years later (Buzas, Lynch, & Berg, 1991). After further work, the team submitted "Regulatory Management and Evaluation of Telephone Service Quality" to *Management Science* in late 1989. This analytical study attempted to bridge principles and regulatory practice. In addition, a review paper on service quality was presented at the Telecommunications Policy Research Conference and the Southern Economic Association meetings that fall. This second working paper was directed to a mix of academics and technically trained policymakers. It reviewed the literature and described the work on the quality index consolidating the 38 characteristics. After being rejected by a theoretical journal, a revised version received favorable reviews at *Telecommunications Policy*, in which it was published in early 1992 (Berg & Lynch, 1992).

5.2 Adapting to Reviews

While still waiting for initial academic reviews of the two manuscripts, the team obtained formal reactions from the FPSC on the "final" report. In a letter, Gene Ferguson, a FPSC engineer, identified a number of deficiencies in the proposed evaluation weighting system:

I would suggest that when the PSC experts are again chosen, they select those of us who understand the rules, procedures, essentials of traffic switching, maintenance and maintenance terms, network structure and network switching, business office and repair facility operation, and the effects of any deficiencies on the subscriber, either directly or indirectly and to the rate payer.

It seems clear the Bureau of Service Evaluation felt that it was not adequately represented in the initial survey! Furthermore, Ferguson was concerned with the wording of the questions and the number of quality dimensions omitted from the initial survey. Lack of initial input from FPSC technical engineers clearly put off the staff. Yet team members thought they were going through correct channels and had no inkling that key stakeholders (in this case, commission engineering staff) had not been utilized, either in developing the questions or taking the survey.

For example, the term *NR* reported in FPSC evaluation reports are often interpreted by companies as *No Rule*. However, the *NR* means that *No* specific percentage *Requirement* was spelled out in the rule. Often 100% compliance is required (when the word *all* is implied). In other cases, the FPSC selects realistic targets for these rules. The initial study did not include these standards.

Also, when staff applied the weighting scheme to a recent evaluation in which a company had failed to meet pass/fail standards in 19 areas, the index score was above average. Although the particular calculation involved a misapplication of the methodology, the example raised red flags regarding the implications of switching to a single index. In addition, in the case of "same-day restoral" (omitted in the initial survey), the FPSC requires 80%. Because this was perceived as an important service standard, they wanted this item included. A number of other omitted items were identified. To illustrate omissions, the team discovered that FPSC engineers had three standards related to central office exchange facilities. In addition, there were seven transmission rules. For example, an exchange with a dial-tone level outside the range of -5 to -22 dBM is not in compliance with the standard. Ferguson wanted some weight given to these items.

A subsequent meeting with FPSC staff in Tallahassee cleared up a number of misunderstandings (on both sides). Follow-up telephone conversations between Buzas and Ferguson helped each understand the others' concerns. The result was a memo by Buzas and Lynch that addressed data collection, aggregation, and evaluation. They raised a number of points.

1. Is the index a gross or net measure of relative benefits? Because service quality is conceptually distinct from cost of service, the focus here was on the benefit side. The trade-off against cost could be handled separately from the relative valuation of quality dimensions. The concern had been expressed that the adoption of a comprehensive index could lead to goldplating. The key point was that for many dimensions, improvements above the standard yields incremental benefits that are greater than the associated incremental costs.[4] Furthermore, the incentives to provide excessive levels of quality were no different than at present.

Just meeting every standard might be more costly than reaching another set of performance targets that yield an identical comprehensive score. If the weights are correct, just meeting each standard is goldplating in the sense that costs are too high for achieving this overall level of quality. Current FPSC mechanisms for evaluating network modernization programs and other prudency tests would apply to the cost side. At least with a single index, the task of the commission (and firms) would be simplified. The FPSC could focus separately on evaluating the additional costs associated with achieving higher quality scores.

2. *Should a firm pass overall when failing several pass/fail standards?* Between 1985 and 1988, companies failed to reach a standard 191 times. Of these, 162 (or 84.8%) were on the 21 (of 38) rules with a standard of 100%. On average, a company failed 13.6 rules of which 11.6 were on rules with standards of 100%. This suggests that the complexity of evaluating performance by 38 (or more) pass/fail standards is problematic when the degree of "substandard" performance is not captured in the summary index. The relative importance of the failures warrants attention.

The fact that a company can fail numerous rules yet receive a "passing" grade is an integral part of the proposed scheme. The research team argued that setting a numerical passing score does not necessitate a degradation in overall quality. Higher overall scores *could* be required (although such ratcheting up ought to be justified in terms of low incremental costs or high incremental benefits). For example, the lowest passing score achieved by any company might be established as the lower bound, or the average passing score could be taken to be the standard. A minimum acceptable performance rule would be established for each rule, whereas overall performance could be gauged on the basis of a higher level than if all such rules were met exactly.

3. *Do the weights reflect a narrow constituency?* We became sensitized to the likelihood that the FPSC had multiple constituencies to protect. Tourists and nonsubscribers rely heavily on public telephones. Residential subscribers are less dependent on public telephones and so would weight less heavily those dimensions associated with public telephone performance. Similarly, businesses that are using digital transmissions would place a premium on other aspects of quality.

It would be possible to develop separate indices for each constituency. Multiple scores could be reported and evaluated, or these could be weighted to obtain an index of aggregate performance. The team did not view such complications as presenting problems. At least the proposed methodology forces decision makers at commissions and companies to acknowledge that different dimensions of quality performance have different weights.

4. *How does the exclusion of important rules from the weighting formula affect its usefulness?* As it turned out, omitted rules tended to be those without set standards for performance. The initial FPSC liaison was asked about this issue prior to developing the initial survey. The liaison was later sent the questionnaire for prior approval. Unfortunately, that person was in the process of leaving the

commission. Lesson: When trying to interact with regulatory agencies, be sure you have the "right" contact point. Academics might not fully appreciate the need to have an onsite advocate for the methodology (or conceptual framework) under consideration. The exclusion of important rules hurt the initial index's credibility.

5. *Did the right experts complete the survey?* Because the initial liaison did not distribute the survey instrument to key technical personnel with knowledge of the standards, the commission weights were brought into question. However, the weights obtained from the telephone company panel of experts were statistically the same as those obtained from the FPSC. It was clear, however, that a new survey had to be developed and given to additional FPSC staff as well as to cooperating telcos.

6. *How were the formulae to be calculated?* As with any new methodology, a clear understanding of its components was essential for its successful application. In the initial FPSC application, incorrect weights were given to several rules. Also, there was a misunderstanding regarding calculation of the overall index. These points could be easily addressed by preparing clearer instructions, including background information and illustrative calculations.

In summary, the exchange of memos and meetings increased the research team's awareness of the administrative processes used to evaluate quality. Furthermore, the official FPSC response to the conceptional framework remained positive. J. Alan Taylor, Chief of the Bureau of Service Evaluation, noted in a letter that he found:

> the analysis to be an excellent and insightful approach to the problems facing regulators who base their quality of service decisions on simple pass/fail rule criteria. Certainly the regulated industry has long been apprehensive of a service quality measurement regime which focuses primarily on failures, without giving some consideration to more economical improvements in overall performance levels. PURC's weighting system therefore appears to be an appropriate way to assure that general quality of service levels remain high, particularly as we move into an era of regulatory flexibility through an incentive approach to governmental oversight.

The team was anxious to address concerns, while directing attention to the advantages to have a comprehensive quality index. The FPSC decided to fund a new survey and the development of new weights. Lynch and Buzas began the new study.

5.3 New Data and Further Regulatory Adaptations

As the new data collection effort proceeded in early 1990, reviews on the comprehensive academic paper were received. Major revisions were required. The most important criticism was that the team should have asked end-users (consumers) what their trade-offs were rather than asking expert regulators and managers to give their views of the trade-offs that would be in consumers' interests. The team had described theoretical reasons to expect that customers of regulated monopolies would not have established trade-offs—because they have no occasion to choose—but reviewers were unconvinced. The team argued that only if they have expertise are consumers likely to have established trade-offs.

Also, the framework had been presented to the academic and national regulatory community in Fall 1989. Eli Noam served as a discussant of the paper at the Telecommunications Policy Research Conference. As a Commissioner for the New York Public Service Commission and as a fellow academic, he supported the framework, but questioned the linearity assumption. He saw no need to have a symmetry of overperformance to underperformance.[5] However, if one more person's call does not go through, it is not clear why the customer's loss depends on whether or not the standard is already exceeded. Nevertheless, it was agreed that more careful empirical investigation of unchanging weights was warranted.

By late Fall 1990, Lynch and Buzas had a draft of the new report, including a LOTUS 1-2-3 spreadsheet that allowed the FPSC to easily calculate a company's overall quality score using the weights uncovered by the project. The Executive Summary was for practitioners. Technical supporting material was included so that an independent expert could revise the weighted index.

The rule clusters in the second study focused on residential service. Besides retaining 16 public telephone rules, additional rules were added, yielding a new set. In Study 2, the weight given to 1 percentage point improvement leading up to the standard was found to be significantly higher than the weight for 1 percentage point improvements exceeding the standard. Two weights were now utilized:

$$W_i^- \text{ if } Z_i < Z_i^*$$
$$W_i^\dagger \text{ if } Z_i > Z_i^*.$$

Thus, the concerns of Florida's PSC staff and New York's Commissioner Noam were supported by the new empirical results. For example, a telephone company had its score for dial-tone delay increased by .111 for each percentage point in excess of the standard, but the score fell by .2133 for each percentage point below the standard. It is possible that technical staff completing the second survey were implicitly factoring in the costs of exceeding standards. In another change from Study 1, technical staff from a major customer of telephone services (Florida's Department of General Services—DGS) completed the new survey instrument, allowing a check on the comparability of FPSC and user weights.

In December 1990, FPSC initiated rule making. The rule-making process turned out to be long and tortuous. Ultimately, Lynch testified on the methodology in mid-1991. However, a glitch appeared that further delayed things. The index got caught in regulatory cross-fire. In February 1991, the Public Counsel's

Office filed complaints that a telco had falsified quality-of-service reports. The company was alleged to have falsified records on out-of-service repairs (a pass/fail quality standard). There are two related standards: 80% for same-day repairs; 95% for within 24 hours. The latter is viewed by some as particularly tight. Corporate officials have denied that they encouraged the falsification of records, although technicians may have felt some pressure to do so.

What actually happened is still under debate (and delaying a rate case). The key point is that the regulatory process can get choked through strategic maneuvering by key stakeholders. The Public Counsel's Office did not like the incentive plan that the FPSC had previously adopted for the telco to begin with. The office was unconvinced that rate payers would benefit from the plan. The falsification report provided a wedge for attacking the plan because quality of service was part of the incentive plan: Rate payers would be due refunds. Staff recommendations were deferred.

Thus, if the FPSC replaced the host of pass/fail standards with a single index, some of the steam would taken out of the Public Counsel's case. The telco might argue "mitigating circumstances" because the old rules had been jettisoned. With tens of millions of dollars at stake, there seemed no reason to replace the many standards with a single quality index. Ironically, had the weighted standard been in effect, the alleged undue pressure to falsify might have been a moot point.

In July 1992, rule making was initiated again; a meeting was scheduled for September 22 on Rule No. 25-4.080 (Weighted Measurement of Quality of Service) with the following FPSC announcement:

> The purpose of this new rule is to introduce another tool which the Commission can utilize in its effort to accurately measure the quality of service provided by local exchange telephone companies.... The rule authorizes the Commission to utilize a weighted index system when considering the adequacy of service.... The system contains various quality of service measures currently contained in Commission rules and weights them according to their importance in the provision of acceptable quality local telephone service.... Companies shall be responsible for complying with each service standard, whether or not an overall score of seventy-five (75) is achieved when the weighted index is employed.

An overall score of 75 corresponded to just meeting each standard. The proposed rule has yet to make it back to the Commissioners for a final vote.[6] If accepted, the new rule will use the weighted index as an additional tool for evaluating quality. In contrast to the research team's recommendations, it will not replace the dozens of pass/fail standards.

One other development substantiated initial telco concerns that the new quality index may be used to tighten standards. The Public Counsel's Office opposed the index because some of the standards with high weights were relatively easy to attain. FPSC staff considered raising dial-tone delay of less than 3 sec from 95% to 98.5% of the time. Similarly, call completion standards might be raised from 95% to 98%. Thus, if the index were adopted, the effective pass/fail standards could also be tightened.[7] Furthermore, some pass/fail standards have not been adopted as formal "rules," but if the Commissioners adopt the index, these standards might become pass/fail rules as well. This development would add to the quality constraints imposed on firms. So the formula and its components may become a bargaining point in the regulatory process. This should have come as no surprise and partly explains why changes in the status quo will be opposed by key stakeholders.

Sample calculations for a telephone company are shown in Table 5.1. In this particular case, the absence of information on directory assistance billing accuracy caused a deduction of over 7 points—but this was more than compensated for by surpassing standards in other categories.

5.4 Academic Revisions

At about the same time as the second rule making was initiated (1992), the reviews on the *Management Science* resubmission arrived. These comments required significant additional research (which was already being undertaken to address issues raised by the FPSC). Different issues arose related to consumer perceptions and valuations. Academic reviewers were still concerned that expert regulators rather than consumers filled out the survey instrument. In anticipation of these concerns, the research team had obtained a sample from the largest purchaser of telephone services within the state government—the Department of General Services (DGS). Thus, these respondents were sophisticated and representative of consumers. The team found close agreement between the weights of experts within the FPSC to those of these other experts.

The initial methodology provided a measure of quality; but if quality improvements lead to higher costs to rate payers, the team was unable to say whether such improvements should be encouraged or discouraged. The team had suggested that this would require regulators to marry the system for measuring the marginal benefits of quality with a detailed study of the marginal costs of improvement along various dimensions. Although the spirit of the weighted index was to get the regulators out of the business of micro-managing regulated companies, it was understood that the cost side warranted attention.[8]

The second addition to the initial study extended the work to address this issue. The revised survey instrument elicited trade-offs between various dimensions of quality and the price of basic monthly service, allowing the team to quantify the dollar value of a 1 percentage point improvement along the various dimensions of service quality. Thus, regulators could encourage a quality improvement that would be coupled with a $1 increase in price if the increase in consumer benefits associated with the improvement exceeds the disutility associated with the price increase. Interestingly enough, the FPSC experts gave a dollar value associated with quality increases that was *three times* that obtained from the large demander (DGS).

TABLE 5.1
WEIGHTED INDEX

ST. JOSEPH TELEPHONE CO. REPORT DATE: APRIL 10, 1992
DATES STUDIED: NOVEMBER 11 THRU DECEMBER 13, 1991

CRITERION	FPSC STANDARD	COMPANY RESULTS	WEIGHT FACTORS	DIFF	WEIGHT ADJUST
A. DIAL-TONE DELAY					
DIAL-TONE DEL +	95.0	100.0	1.1377	5	5.6884
DIAL-TONE DEL –	95.0		8.4935		
B. CALL COMPLETIONS					
INTRAOFFICE +	95.0	100.0	0.0613	5	0.3063
INTRAOFFICE –	95.0		4.0136		
INTEROFFICE +	95.0	100.0	0.0947	5	0.4735
INTEROFFICE –	95.0		2.1075		
EAS +	95.0	100.0	0.0280	5	0.1402
EAS –	95.0		0.9953		
INTRALATA DDD +	95.0	99.9	0.1286	4.9	0.6300
INTRALATA DDD –	95.0		1.0999		
C. INCORRECTLY DIALED CALLS					
INCORRECTLY DIALED +	95.0	100.0	0.1043	5	0.5214
INCORRECTLY DIALED –	95.0		0.1043		
D. 911 SERVICE					
911 SERVICE –	100.0	100.0	2.8772		
E. TRANSMISSION					
DIAL-TONE LEVEL –	100.0	100.0	0.0002		
CENTRAL OFFICE LOSS –	100.0	100.0	0.0002		
M.W. FREQUENCY –	100.0	100.0	0.0002		
CEN. OFF. NOISE METAL –	100.0	50.0	0.0002	–50	–0.0118
CEN. OFF. NOISE IMPLSE –	100.0	87.8	0.0002	–12.2	–0.0029
SUBSCRIBER LOOPS +	98.0	100.0	0.2788	2	0.5577
SUBSCRIBER LOOPS –	98.0		0.1394		
F. POWER AND GENERATORS					
POWER & GENERATORS –	100.0	100.0	0.0798		
G. TEST NUMBERS					
TEST NUMBERS –	100.0	100.0	0.0010		
H. CENTRAL OFFICE					
SCHEDULED ROUTINE PROG +	95.0	100.0	0.0487	5	0.2433
SCHEDULED ROUTINE PROG –	95.0		0.0487		
FRAME +	95.0	100.0	0.0549	5	0.2743
FRAME –	95.0		0.0549		
FACILITIES +	95.0	100.0	0.0758	5	0.3790
FACILITIES –	95.0		0.0758		
I. ANSWER TIME					
OPERATOR +	90.0	99.1	0.0519	9.1	0.4725
OPERATOR –	90.0		0.3820		
DIRECTORY ASSISTANCE +	90.0	97.0	0.0519	7	0.3635
DIRECTORY ASSISTANCE –	90.0		0.3820		
REPAIR SERVICE +	90.0	100.0	0.0519	10	0.5192
REPAIR SERVICE –	90.0		0.3820		
BUSINESS OFFICE +	80.0	98.2	0.0604	18.2	1.0994
BUSINESS OFFICE –	80.0		0.4191		

(Continued)

TABLE 5.1
(Continued)

ST. JOSEPH TELEPHONE CO. REPORT DATE: APRIL 10, 1992
DATES STUDIED: NOVEMBER 11 THRU DECEMBER 13, 1991

CRITERION	FPSC STANDARD	COMPANY RESULTS	WEIGHT FACTORS	DIFF	WEIGHT ADJUST
J. ADEQUACY OF DIR. AND DIR. ASSISTANCE					
DIRECTORY SERVICE –	100.0	100.0	0.0887		
NEW NUMBERS –	100.0	100.0	0.0399		
NUMBERS IN DIRECTORY +	99.0	100.0	0.2507	1	0.2507
NUMBERS IN DIRECTORY –	99.0		0.5640		
K. ADEQUACY OF INTERCEPT SERVICES					
CHANGED NUMBERS +	90.0	100.0	0.1287	10	1.2865
CHANGED NUMBERS –	90.0		0.3107		
DISCONNECTED SERVICE +	80.0	100.0	0.0489	20	0.9775
DISCONNECTED SERVICE –	80.0		0.2151		
VACATION DISCONNECTS +	80.0	100.0	0.0322	20	0.6434
VACATION DISCONNECTS –	80.0		0.0586		
VACANT NUMBERS +	80.0		0.0277		
VACANT NUMBERS –	80.0		0.2079		
DISCONNECTS NON-PAY –	100.0		0.1650		
L. TOLL TIMING AND BILLING ACCURACY					
INTRALATA BILL ACC. +	97.0	100.0	0.4290	3	1.2869
INTRALATA BILL ACC –	97.0		2.8560		
DIR. ASSIST. BILL ACC. +	97.0		0.4794		
DIR. ASSIST. BILL ACC. –	97.0	0.0	0.0766	–97	–7.4277
M. PUBLIC TELEPHONE SERVICE					
1 PAY PHONE/EXCHANGE –	100.0	100.0	0.0006		
SERVICEABILITY –	100.0	92.0	0.0864	–8	–0.6910
HANDICAPPED ACCESS –	100.0	96.0	0.0112	–4	–0.0449
GLASS +	95.0	100.0	0.0056	5	0.0278
GLASS –	95.0		0.0056		
DOORS +	95.0	100.0	0.0051	5	0.0254
DOORS –	95.0		0.0051		
LEVEL +	95.0	100.0	0.0076	5	0.0379
LEVEL –	95.0		0.0062		
WIRING +	95.0	100.0	0.0060	5	0.0298
WIRING –	95.0		0.0141		
CLEANLINESS +	95.0	100.0	0.0005	5	0.0024
CLEANLINESS –	95.0		0.0362		
LIGHTS –	100.0	92.0	0.0224	–8	–0.1793
TELEPHONE NUMBERS –	100.0	100.0	0.0523		
NAME OR LOGO –	100.0	100.0	0.0008		
DIAL INSTRUCTIONS –	100.0	92.0	0.0864	–8	–0.6910
TRANSMISSION +	95.0	96.0	0.0266	1	0.0266
TRANSMISSION –	95.0		0.0266		

(Continued)

TABLE 5.1
(Continued)

ST. JOSEPH TELEPHONE CO. REPORT DATE: APRIL 10, 1992
DATES STUDIED: NOVEMBER 11 THRU DECEMBER 13, 1991

CRITERION	FPSC STANDARD	COMPANY RESULTS	WEIGHT FACTORS	DIFF	WEIGHT ADJUST
DIALING +	95.0	100.0	0.0008	5	0.0040
DIALING −	95.0		0.0062		
COIN RETURN AUTO −	100.0	96.0	0.0037	−4	−0.0147
COIN RETURN OPER +	95.0	96.0	0.0178	1	0.0178
COIN RETURN OPER −	95.0		0.0178		
OPERATOR ID COINS +	95.0	96.0	0.0002	1	0.0002
OPERATOR ID COINS −	95.0		0.0302		
ACCESS ALL LD CARRIERS −	100.0	100.0	0.0024		
RING BACK OPERATOR +	95.0		0.0002		
RING BACK OPERATOR −	95.0	92.0	0.0302	−3	−0.0905
COIN FREE ACCESS OPER −	100.0	100.0	0.0097		
COIN FREE ACCESS D.A. −	100.0	100.0	0.0042		
COIN FREE ACCESS 911 −	100.0	100.0	0.0093		
COIN FREE ACCESS R.S. −	100.0	100.0	0.0034		
COIN FREE ACCESS B.O. −	100.0	100.0	0.0027		
DIRECTORY −	100.0	100.0	0.0013		
DIRECTORY SECURITY +	95.0	100.0	0.0510	5	0.2551
DIRECTORY SECURITY −	95.0		0.0510		
ADDRESS/LOCATION −	100.0	92.0	0.1252	−8	−1.0013
N. AVAILABILITY OF SERVICE					
3-DAY PRIMARY SERVICE +	90.0	100.0	0.0333	10	0.3332
3-DAY PRIMARY SERVICE −	90.0		0.2406		
PRIM. SERV. APPOINTMNT +	95.0	100.0	0.1306	5	0.6528
PRIM. SERV. APPOINTMNT −	95.0		0.8125		
M. REPAIR SERVICE					
RESTORED-SAME DAY +	80.0		0.0909		
RESTORED-SAME DAY −	80.0	71.2	0.1319	−8.8	−1.1603
RESTORED-24 HOUR +	95.0	100.0	0.3685	5	1.8427
RESTORED-24 HOUR −	95.0		1.3348		
REPAIR APPOINTMENTS +	95.0	96.8	0.1318	1.8	0.2372
REPAIR APPOINTMENTS −	95.0		0.1936		
REBATES OVER 24 HOURS −	100.0	100.0	0.0523		
SERVICE AFFECTING-72 HRS +	95.0	100.0	0.1318	5	0.6590
SERVICE AFFECTING-72 HRS −	95.0		0.1936		
P. CUSTOMER COMPLAINTS	ST. AVE				
COMPLAINTS/1000 LINES +	0.42	0.36	0.3685	0.1333	0.0491
COMPLAINTS/1000 LINES −	0.42		0.0000		
BASE SCORE IF ALL					
STANDARDS ARE MET					
EXACTLY			75.00		75.00
SUM OF ADJUSTMENTS					9.00
OVERALL WEIGHTED SCORE					
(BASE + SUM OF					
ADJUSTMENTS)					84.00

The average for the FPSC and DGS implied that a \$1 change in price was worth .1401 points.[9] If performance on the dial-tone delay standard improved from 97% to 98%, the associated change in the point score was .111 (on a 10-point scale). Thus, a 1 percentage point improvement is worth .111/.1401/\$ = \$0.79 per month to the customer. This application of the framework to benefit cost analysis represents a potentially valuable extension of this methodology.

5.5 Customer Perceptions and Expert Trade-Offs

It is useful to underscore the role of experts in the team's framework. The second study and associated revisions of academic papers attempted to address the key conceptual issues raised by reviewers.[10] When expert regulators assess service quality in the interest of everyday consumers, the tacit assumption is that the experts' utility functions are highly correlated with those that everyday customers would have if they did not lack knowledge of how measurable attributes translate into realized benefits. Another justification for using experts is that research indicates that when trade-offs among attributes do not already exist in respondents' heads, the trade-offs they construct on the spot are highly unstable. There is a substantial literature showing that experts make trade-offs that are much less sensitive to distorting effects of measurement. The relevant issue is not whether the consumer has experience with the product. The critical concern is whether consumers have experience making trade-offs among the particular dimensions relevant to the decision at hand. Experience means that these trade-offs can be retrieved rather than constructed at the time of measurement (Feldman & Lynch, 1988; Fischhoff, 1993).

The team's approach is supported by Fischhoff (1993), who surveyed a large body of applied policy research that uses "contingent valuation" methodology. That research attempts to elicit citizens' values in the hope that spending on public policy programs can reflect the priorities of consumers. The repeated finding is that when consumers' trade-offs are elicited for goods that are not customarily traded in any marketplace, consumers do not have articulated values relevant to those decisions. The result is that measured trade-offs have indefensible properties.

One could still argue that consumers should identify the criteria to be measured, and that the role of regulators might be circumscribed to judging the quality of credence attributes—those characteristics that are rarely learned, even after consumption. In fact, this process is very close to what actually happens. Consumers call in their complaints to companies and regulators. Such data are reported. The complaints are invariably phrased in terms of benefits. Legal constraints force regulators to reverse engineer those complaints and to trace them back to problems on attributes that underlie those benefits. The problem is that when laws and rules have been written pertaining to these objective attributes (e.g., percentage of interoffice calls completed), they become credence attributes for consumers. Typical consumers do not understand the links between technical attributes and benefits.

When comparing expert regulators and novice consumers, it is expected that the former generate weights that would be strongly related to those elicited from a trained representative sample of consumers. Consistency across companies, the FPSC, and a large state agency that buys telecommunications services support the research team's view that the weights provide a good first cut at prioritizing the dimensions of quality.[11]

6. CONCLUDING OBSERVATIONS

What was learned? The regulatory process is run by lawyers, with the aid of accountants and engineers. Economists have input into the process, but the deference is underwhelming. Perhaps, this is appropriate. After all, as a profession, precious little has been developed that helps decision makers identify and reward quality.

The research team was pleased that the multiyear research project resulted in a proposed rule for adopting the weighted index system for the evaluation of service quality. However, the proposed rule utilizes the index as an added requirement rather than as a replacement indicator of pass/fail performance standards. Because the spirit of the recommendation was to move away from detailed consideration of the 60-plus dimensions of quality, it is hoped that the FPSC ultimately adopts the comprehensive performance index. If the FPSC utilizes the crude, unweighted pass/fail mechanism as well as the comprehensive index during the transition period, that is their judgement call as to what is politically acceptable.

Are there other lessons from this case study? The multiyear project represents an attempt to serve two masters: one interested in implementing regulatory policy and another interested in extending the boundaries of science. Table 5.2 lists some of the lessons learned in the process. Regulatory review and academic review have similar properties. Each has well-established criteria, although the weights given each will differ dramatically. Academics put a premium on elegance (although simplicity sometimes wins the day). Certainly, regulators will emphasize simplicity over complexity. Both seek robustness of results. The conclusions need to stand up to possible changes in initial conditions. Academics place a premium on the new and innovative, whereas regulators emphasize continuity. After all, telecommunications infrastructure assets are so long-lived that switching policies can wreak havoc with decision making. But there is also an element of continuity that academics do respect. The accepted paradigm will not be easily displaced, so the policy conclusions had better square with prior views with regards to the setting. It is okay for the results to be counterintuitive, so long as they are based on maximizing behavior by individual (generally, well-

TABLE 5.2
Telephone Service Quality Analysis: Lessons Learned
From Serving Two Masters

1. The conceptual frameworks provided by economics and other decision sciences can shed light on extremely complicated phenomena.
2. The application of analytical and empirical tools to real-world problems requires a solid understanding of both the tools and the context in which they are being applied.
3. It is important to maintain close interactions with policymakers, checking to make sure that other stakeholders are not excluded from the process.
4. Present preliminary results in a variety of academic, corporate, and regulatory forums in order to benefit from expertise that exists outside your own research team.
5. Real policy change takes time—to obtain feedback, revise analyses, and convince decision makers of the merits of the change. Nothing is guaranteed.
6. Similarly, scientific review is a time-intensive process designed to (a) identify truly innovative and insightful ideas, and (b) screen out those that do not meet scholarly standards.
7. No one said it would be easy. If untenured, do not attempt to serve two masters!

informed) agents! Even more problematic for *Management Science* reviewers has been the team's perceived disregard for the marketing paradigm: "Quality is what customers perceive it to be." The team's efforts at explaining the role of experts in the regulatory process have (so far) been only partially successful.

Both processes are designed to kill bad or useless ideas. Thus, review lags are not only reasonable, but necessary if the contribution is to be evaluated carefully and thoughtfully. Rejections by an editor involve some randomness: The particular reviewer does not really understand the paper (alternatively, the points are not expressed in a logical and careful manner), the reviewer has an irrational grudge against a line of research (my favorite excuse), or the reviewer is on sabbatical and the disorganized editor lets one languish in purgatory for an unseemly amount of time. Whatever the reasons for the lag and rejection, the process is honored and is widely believed to improve the contributions to the scientific literature.

Good reviewers provide detailed feedback when the paper shows promise. Initial versions of papers are seldom ready for prime time. A parallel process occurs in the adversarial regulatory setting. Various stakeholders will identify limitations to the proposed policy. Alternatively, one group might stonewall an idea if implementation would be injurious to its position. The art of bargaining and compromise are probably better developed in the regulatory arena than in academia. The pursuit of "truth" can get in the way of "pretty good" policies.

So this tale still has no ending. The most comprehensive (and rigorous) expression of these ideas is still under review at a highly ranked journal. The feedback has been thorough and the team has tried to be responsive to reviewer suggestions. Similarly, the team still waits for Commission passage of a rule that just adds the proposed index to the regulatory toolkit for evaluating performance. In both cases, the lags seem long—but they are also understandable, given the stakes in both instances.

ADDENDUM

One week after the CITI Conference, the *Management Science* paper was accepted subject to minor revisions (see Lynch, Buzas, & Berg, 1994). The use of experts instead of consumers in developing weights no longer blocked academic acceptance of the conceptual framework. On June 14, 1993, the proposed FPSC rules became effective, so the index is now part of the regulatory process in Florida! The dual missions *were* accomplished.

ACKNOWLEDGMENTS

Helpful comments from John Lynch, Alan Taylor, Bill Lehr, and Jill Butler are acknowledged (without implicating them).

APPENDIX[12]

Hierarchical conjoint analysis circumvents problems with more commonly used variants of conjoint analysis: full profile analysis and two-at-a-time trade-offs. In the former, judges would have to keep information of all the scores in their minds—integrating them into an overall evaluation. Two-at-a-time trade-offs require the respondents to fill out matrices in which they evaluate profiles of combinations of high and low levels of performance of all possible pairs of dimensions. If there are 38 different dimensions, this involves filling out 703 different matrices.

The weighting scheme was developed by having experts from the FPSC and telcos rate different hypothetical company profiles of performance. Table 5.3 depicts one expert's ratings for four hypothetical combinations. Possible scores ranged from 1 (worst possible overall performance) to 10 (best possible overall performance). The observable range for high and low performance for each standard was established with the assistance of FPSC staff. Here, two aspects of service availability matter: (a) restoring primary service within three days of an outage, and (b) keeping appointments the morning or afternoon for which they are scheduled. Service restoration of 99% and meeting appointments 100% was rated 10 by this expert judge. If service restoration dropped to 88%, the combination was scored as an 8. Scores drop in a similar fashion if appointments are kept at the lower level, whereas restoration drops from 99% to 88%.

Thus, within the service availability cluster, one could distinguish between high and low levels of performance. When performance on the appointments standard is highest, the average score is 9((10 + 8)/2). Whereas when performance on the appointments standard is lowest, the average score is 5((6 + 4)/2). Thus, to calculate the weights within the service availability cluster, one can note the

TABLE 5.3
Rating Company Profiles

(a) Availability of Service			
		Appointments = 100%	Appointments = 94%
3-Day Primary Service	99%	10	6
	88%	8	4

(b) Calculation of Effects on Ratings within Cluster			
Dimensions	Average for High Level	Average for Low Level	Difference Between Averages
Appointments	(10 + 8)/2 = 9	(6 + 4)/2 = 5	4
3-Day Primary Service	(10 + 6)/2 = 8	(8 + 4)/2 = 6	2

(c) Calculation of Weights within Cluster			
Dimensions	Difference Between Averages	% Spanned	Weight Within Cluster
Appointments	4	100 − 94 = 6	.667
3-Day Primary Service	2	99 − 88 = 11	.222

difference between these averages of 4. The number of percentage points spanned between high- and low-appointment performance is 4 performance points over the span of 6 percentage points. Thus, each 1 percentage point improvement in appointments implies a .667 increase in the score (on a scale of 1 to 10).

Note that the weights are the result of numbers assigned by evaluators (representing informed consumers). One expert respondent might assign a combination a 6 for a 3-day primary service and appointments of 99% and 94%, respectively. Another might assign a 7. The initial comparisons were ordinal in the sense that they are numerical representations of preference orderings. However, they have been treated as cardinal for policy purposes because a dollar metric was introduced in the second study that nailed the numbers to a dollar value. In addition, the numerical rankings across experts were statistically quite similar.

Table 5.4 illustrates the kinds of tests used to determine agreement among experts. The weights from four experts are shown. In this example, there is high disagreement regarding 3-day primary service. The sample mean (based on two observations from Experts 1 and 2) departs from the "true" weight that would be derived from all four experts. However, the data give a warning signal in that the sample standard deviation is high. For attributes in which there is low

TABLE 5.4
Weight of 1 Percentage Point Changes

Expert	Appointments	3-Day Primary Service
1	.200	.167
2	.220	.500
3	.180	.100
4	200	.400
Population Mean of 4	.200	.292
Population Standard Deviation	.014	.164
Mean Based on 1 & 2	.210	.334
Standard Deviation Based on 1 & 2	.014	.235

disagreement—appointments—the sample mean will be close to the "true" weight that would be derived from all four experts (here, the population). The low standard deviation captures this feature of the sample.

Similar respondent rankings for comparisons across clusters allowed the derivation of a comprehensive scoring formula containing weights for each 1-percentage point improvement in the different telephone service quality dimensions. The results allow one to have some confidence in the identification of quality dimensions that really matter and in the relative weights to be given those dimensions. There is always room for refinements, but the methodology represents a major improvement over pass–fail approaches now utilized by regulatory commissions.

ENDNOTES

1. Edwards (1977) found that when a series of multidimensional land-use proposals were evaluated intuitively and holistically by regulators and developers, the parties differed substantially in their rank orders of the proposals' desirability. When the relative values the parties placed on the dimensions were measured and modeled, the evaluations derived from the models of all parties exhibited remarkable agreement. Edwards suggested that the disagreements in holistic judgments stem from unconscious tendencies to focus on the subset of dimensional cues that are consistent with the overall judgment the parties would like to draw. In the present context, such selective processing can lead company representatives to truly believe that their companies provide superior quality, whereas skeptical regulators reach a contrary conclusion from the same evidence.

2. Note that the scoring function presented earlier is merely a transformation of the constraint equation described here:

$$Q_a = W_0 + W_1(Z_1 - Z_1^*) + W_2(Z_2 - Z_2^*)$$
$$Q_a - (W_0 - W_1 Z_1^* - W_2 Z_2^*) = W_1 Z_1 + W_2 Z_2$$
$$Q = W_1 Z_1 + W_2 Z_2$$

3. Whitley (1991) argues that "The role of journals in economics is closely connected to the nature of economics as a scientific field . . . journals dominate the formal communication system,

are ordered into a strong prestige hierarchy, reproduce a strong analytical orthodoxy and publish highly formal, standardized and concise papers" (p. 32).

4. Given the difficulties in sorting out potential cost complementarities, the quality dimensions might be considered in related bundles.

5. Noam (1991) addressed numerous implementation issues, including the integration of service quality indices into the incentive process. For example, "Gold-plating could . . . be dealt with by setting ceiling for rewards" (p. 185).

6. The use of 75 as the base score for just meeting each standard involved a simple transformation of the calculated coefficients, although it complicates comparisons with the initial study (of 38 dimensions).

7. An alternative, methodologically cleaner, approach would be to raise the passing score to 80 or 85, in which 75 represented just meeting all initial standards.

8. As has already been noted, the differential weight for sub- and superperformance was a feature of the second study that might be capturing a regulatory concern for goldplating.

9. The derivation is available in Lynch et al. (1994). Note that the proposed rule utilizes weights for a 100-point scale, in which 75 is the score for just satisfying each rule.

10. I am indebted to John Lynch for formulating the points related to customer versus expert perceptions.

11. A similar approach was recommended to the New York Public Service Commission in a Theodore Barry & Associate (TB&A) study on performance improvement opportunities at the New York Telephone (NYT) Company (Mayer, 1993). Their service quality index had seven elements: dial line, customer contact, maintenance, installation, customer expectation, and operational efficiency. The TB&A study evaluated NYT programs in terms of impacts on these categories. It attempted to identify trade-offs among construction, maintenance expense levels, and service quality. As such, the approach provides both a management tool (for investment planning) and a regulatory analytic technique (for evaluating corporate performance). The TB&A line of investigation illustrates the increased attention being given to indexes and weights for addressing telephone service quality issues.

12. Thanks to Tom Buzas for developing the example used in the Appendix and in Table 5.3, and to John Lynch for the example in Table 5.4.

REFERENCES

Berg, Sanford V., & John G. Lynch, Jr. (1992). The measurement and encouragement of telephone service quality. *Telecommunications Policy, (April)*, 210–224.

Buzas, Thomas E., & John G. Lynch, Jr. (1988). *A formula for the comprehensive evaluation of local telephone companies: Report to the Florida Public Service Commission: Preliminary report.* (January).

Buzas, Thomas E., John G. Lynch, Jr., & Sanford V. Berg (1991). Issues in the measurement of telephone service quality. In Barry Cole (Ed.), *After the breakup: Assessing the new post AT&T divestiture era* (pp. 268–276). New York: Columbia University Press.

Edwards, E. (1977). How to use multiattribute utility measurement for social decision making. *IEEE Transactions in Systems, Man and Cybernetics, 7*, 326–340.

Feldman, J. M., & John G. Lynch, Jr. (1988). Self-generated validity and other effects of measurement on belief, attitude, intention, and behavior. *Journal of Applied Psychology, 73*, 421–435.

Fischhoff, Baruch (1993). Value elicitation: Is there anything in there? In M. Hechter, L. Nadel and R. Michod (Eds.), *The Origin of Values* (pp. 187–214). Hawthorne, NY: Aldine de Gruyter.

Louviere, Jordan (1984). Hierarchical information integration: A new method for the design and the analysis of complex multiattribute judgement problems. In Thomas Kinnear (Ed.), *Advances in consumer research, Vol. 11* (pp. 148–155). Provo, UT: Association for Consumer Research.

Lynch, John G. Jr., Thomas E. Buzas, & Sanford V. Berg (1994). Regulatory measurement and evaluation of telephone service quality. *Management Science, February,* 169–194.

Mayer, Robert H. (1993). *Integrating service quality, customer service and alternative investment analysis.* TB&A Management Consultants, April 23.

Noam, Eli M. (1991). The quality of regulation in regulating quality: A proposal for an integrated incentive approach to telephone service performance. In M. A. Einhorn (Ed.), *Price caps and incentive regulation in telecommunications* (pp. 167–189). Boston: Kluwer Academic Publishers.

Rovizzi, Laura, & David Thompson (1992). The regulation of product quality in the public utilities and the citizen's charter. *Fiscal Studies, 13:3,* 74–95.

Whitley, Richard (1991). The organization and role of journals in economics and other fields. *Economic Notes, 20:1,* 6–32.

Chapter 6

Network Utilization Principles and Pricing Strategies for Network Reliability

Raymond W. Lawton
The National Regulatory Research Institute
The Ohio State University

1. INTRODUCTION

Regulatory and public policymakers have become increasingly concerned about the current and future reliability of the public-switched telecommunications network. Part of this concern is driven by the worry that the cost-efficiency incentives contained in price caps and other regulation reforms may cause networks to reduce their costs in ways that also reduce the reliability of the network. A second source of concern is more immediate and is sparked by the number of recent and widely reported outages affecting major U.S. telecommunications carriers. The third is the interest in ensuring that a modern communications infrastructure exists that can reliability serve the new and emerging needs of an information-age economy. However, reliability issues have not enjoyed the same level of attention that pricing, costing, and market structure have received from regulators, policymakers, and academic researchers. Accordingly, a widespread consensus on how to define, measure, price, or resolve network reliability is lacking.

Complicating the reliability issue even further is the notion that although networks are widely used, they are not fully understood. The airline, trucking, telecommunications, and computer industries are visible examples of the pervasive and successful use of networks in modern economies. In each of these industries the design, management, operating characteristics, and output of the networks has been extensively studied.[1] Yet, the economic rationale of these networks—namely, why networks are used versus other modes of production—

135

has not been adequately examined. This shortcoming is particularly significant because it is the economic advantage provided by a network as well as its reliability and other operating characteristics that account for the successful use of networks in modern economies.

It is argued here that costing and pricing approaches used by state and federal utility regulatory commissions that appear to work well for some types of networks do not seem to be as well suited for telecommunications networks. Commission-approved pricing based on incremental costs, for example, has produced visibly efficient outcomes in the electric utility networks because of their increasing-cost curve, but it does not seem to be as efficient for the declining costs of public-switched telecommunications networks. Costing and pricing schemes are generally unsuccessful when they do not accurately reflect the underlying cost structures as well as the pricing environment of the networks. If networks can be viewed as being in either regulated, transitional, or competitive markets, then the optimal commission-authorized networks' costing and pricing approach to be used for each specific type of market is the one that is also congruent with the underlying cost structure. A simple example of this would be the inappropriateness of a regulatory commission using the average cost pricing long associated with regulated telecommunications markets for a competitive telecommunications market.

In the following section certain network utilization principles are developed based on the costing and pricing characteristics of telecommunications networks in competitive markets. These network utilization principles are used to describe costing and pricing approaches that are appropriate for certain kinds of network market environments. Use of these principles allows a regulatory commission, a utility, and firms in a telecommunications market to design efficient pricing schemes. Because reliability is a key cost factor and has measurable parameters, these network utilization principles can be applied to assist network owners or managers in designing reliability pricing schemes. Accordingly, the third section of this chapter uses these principles to identify reliability pricing options that regulatory commissions can use for competitive, transitional, and regulated telecommunications networks.[2]

2. NETWORK UTILIZATION PRINCIPLES

The fundamental assumption underlying network utilization principles described here is that the construction, operation, and use of any network is based on economically rational decisions. That is, based on the information available to a firm at any point in time, a network is the least expensive means of delivering to the firm's customers a specific telecommunications service or set of services. If a network is not the least expensive means of delivering a particular service, then an alternate means of delivering the service will be used or established by the firm. The assumption of economic rationality is key to understanding both the purpose served by and the utilization of a network. Unless other reasons exist that override the firm's interests in profit maximization, a network must be the low-cost service provider or it will not be used.

A network may be either a physically interconnected ubiquitous distribution system or an integrated system of switches or nodes and routes or channels with usage restrictions and enforceable interconnection agreements. The local distribution arrangement of electric, gas, and water utilities are generally thought to be the classic examples of the first definition, whereas the public-switched telecommunications network, Peter Huber's vision of a geodesic telecommunications network,[3] and various intermodal transportation systems are examples of the second. More specifically, a telecommunications network may be defined as a ubiquitous and economically efficient set of switched communications flows.

Ubiquity is one of two indispensable components of a network. An information or communications distribution system that does not have the facilities, or instant access to the facilities, needed to serve all customers desiring service within an area is neither ubiquitous nor a network.[4] A telecommunications firm able only to serve a fraction of the customers within an area would not be considered ubiquitous or a network. On the other hand, a telecommunications company that provides telecommunications services to all customers within a building would be considered as having a ubiquitous network for a particular service or set of services when the building is chosen as the unit of analysis. Ubiquity, not size, is one key factor when defining a network.

In a regulatory setting ubiquity is important because a utility is awarded a franchise and must provide service on demand and for a commission-approved price to any consumer within the geographical boundaries of its service territory. For a regulated utility ubiquity is both an obligation and a compelling economic fact of life. If significant economies of scale or scope exist, then the utility may be both the low-cost provider and the only ubiquitous provider. This combination gives the incumbent-regulated telecommunications utility a significant commission-supported advantage over both ubiquitous and nonubiquitous challengers. As telecommunications markets evolve toward being competitive, regulators and policymakers need to better understand the often overlooked concept of ubiquity in order to design optimal, transitional, pricing strategies for currently regulated telecommunications markets.

An entity wishing only to link certain points or customers together is not ubiquitous and is best thought of as a point-to-point network, or subnetwork. Many examples of subnetworks exist. Railroads, for example, have developed extensive private telecommunications systems that allow switched communications flows between any railroad company facility without having to use the public-switched network. However large the resulting railroad telecommunications system, such a system is not ubiquitous as it does not serve (or intend to serve) all potential customers within a geographical area.[5] A local area network (LAN) or metropolitan area network (MAN) or token ring provider that chooses only to serve a particular market segment within a geographical area would also be a subnetwork.

In understanding network utilization the type of ownership and form of regulation are not as important as knowing that a network provides ubiquitous service. Once the major pattern of network utilization has been established, then the ownership and regulatory status of the network become significant factors affecting network utilization.

From both an economic perspective and a public policy viewpoint, both ubiquitous networks and subnetworks are important. Even with changes in technology, market structure, perceived and actual customer demands, and regulatory policy, the future telecommunications network will include both types of networks. To the extent that the point-to-point subnetworks do not desire, need, or demand ubiquity, they should continue to exist as viable nonubiquitous communications systems. No second-class status is inferred by this classification or presumption that interconnection to a network is required, only that subnetworks are not ubiquitous and that this characteristic has pricing and competitive consequences.

Ubiquity suggests another aspect of networks; namely, that they are indivisible in terms of facilities and availability of communications services. Unlike a point-to-point subnetwork, a network, whether or not it has franchise obligations, must offer the uniform availability of a standard and reliable level of communications services for all customers in order for it to be ubiquitous. This point is developed more fully later.

The second indispensable part of a network is the economic advantage the network offers the firm using the network over other networks and over alternative modes of delivering telecommunications services. A firm will use a particular network if it is less expensive for a given level of reliability, quality, and type of service than the choices available from other network, subnetwork, or nonnetwork providers. Based on these underlying network characteristics, six network utilization principles can be stated. The network utilization principles identified can be used for any type of network or network service such as reliability and are elaborated here using the provisioning of switched telecommunications services in a specified geographical area.

Network Utilization Principle 1: A network will only be used by a firm if it is the least costly alternative for the delivery of a particular service or set of telecommunications services.

The first principle directly and immediately follows from the economic advantage characteristic inherent in all successful networks. A firm using a telecommunications network explicitly does so because, for example, it has determined that a targeted network-provisioned, fax-based sales campaign is less costly and more effective than other alternatives such as mass mailings, radio or television commercials, newspaper advertisements, or electronic bulletin boards. Using the particular telecommunications network selected gives the firm an economic advantage. Unless use of the network is restricted, this economic advantage is equally available commercially to all similar firms.

> *Network Utilization Principle 2: A firm will build, rent, or otherwise obtain its own facilities-based network when doing so is less costly than the use of existing commercially available networks.*

The first two principles are founded on the notion that an economically rational firm will choose the least costly means for the reliable delivery of its telecommunications-dependent services to its customers. Economics and engineering economics texts recognize this in their treatment of capital investment decision making by firms. As firms are interested in maximizing future revenue streams, they are indifferent as to whose facility provides them with the needed service. They care only that the facility or network chosen is the least costly alternative for a certain quantity, type, level of reliability, and quality of service.

A firm is, therefore, willing to consider all reasonable alternatives: using an existing network, using one or more point-to-point subnetworks, constructing its own network, or using nonnetwork providers. The bottom line here is that the alternative chosen is the least costly of all those available to the firm that will allow the firm to reliably deliver certain services to its customers. Accordingly, if a particular network is not the lowest cost means of delivering ubiquitous service, then a second network will be constructed, and all customers of that service will prefer the second network because of its lower cost of, for example, reliability. Therefore, the economically successful existence of a network is proof of principle 3.[6]

> *Network Utilization Principle 3: A network is the least cost alternative for the ubiquitous delivery of certain telecommunications services.*

This principle is derived from the twin notions that only networks can provide ubiquitous service, and that a firm will only use a network if it is the least costly option. A geographical area may have more than one network, and each network could have essentially identical costs.[7] From an engineering perspective, switched telecommunications networks gain their efficiency by determining the optimal configuration of switches and lines needed for the telecommunications demand pattern existing for a self-selected service area.[8] By being ubiquitous, a network has a larger volume of traffic than a subnetwork and can better design economically efficient network facilities.[9] Said another way, whereas the network manager does not know if a particular firm will use a specific service at a given time and location, the network manager does know the basic underlying aggregate demand pattern and can build facilities that can handle the demand within a known margin of error. The multidirectional star typology of public-switched, regulated networks, for instance, is ideal for efficiently and reliably handling large volumes of telecommunications traffic because the trunk lines that link switches can be used for multiple purposes. This and other engineering optimization techniques, such as the recent development of "self-healing" fiber-ring typologies, help make telecommunication networks efficient providers. It is important to note that networks are not always the least cost alternative, only that use of a network by a firm (based on the information available) means a firm has selected a particular network as the lowest cost means of delivering or receiving certain services.[10]

Network Utilization Principle 4: A point-to-point network or subnetwork is the least cost alternative for the nonubiquitous delivery of certain telecommunications services.

An economically rational actor will not build or use a subnetwork unless doing so will increase its net revenues and its profits. In general, a point-to-point subnetwork is viable because it has a different underlying economic structure than that of the ubiquitous network. Three important elements describing this difference are traffic concentration, the cost of ubiquity, and the nonubiquitous nature of the services to be sold. In consideration of the first point, many networks have a channel or set of channels that carry a disproportionate share of the traffic of the network. Unless given price discounts, customers of a network will consider the construction or rental of their own subnetwork if they do not need ubiquitous service and instead need only the high-volume routes of the incumbent network. A firm will look for a subnetwork if the change will minimize costs, maintain the same level of reliability, and the users will not require ubiquitous service. Retail store chains with their own internal telecommunications systems are good examples of successful point-to-point subnetworks, although they still depend on a ubiquitous public-switched network for their external calls.

Second, although the primary advantage of a network is its ubiquity, its greatest structural weakness—namely, the presence of localized cost diseconomies—is due to the ubiquitous nature of the network. Just as significant economies are realized by the strategic inclusion of high-traffic routes into a network, substantial diseconomies occur when all customers in a geographical area are linked by the network.[11] The net result of these diseconomies is to provide incentives to the firm needing only to use certain routes to consider the development or use of a subnetwork that minimizes its communications costs.[12]

Third, a point-to-point subnetwork does not, by definition, provide by itself ubiquitous service.[13] Unless a firm never used the ubiquitous network for anything other than the specialized services now provided by the subnetwork, some connection will likely be maintained with a network. However, to the extent that a nonubiquitous service can be provided more economically than had been provided by the network, a point-to-point subnetwork will be preferred.

Networks and subnetworks can and do coexist and prosper when each fills a different need. All other things being equal, subnetworks can only come into being when a network is not the lowest cost alternative for a firm. Again, the economic existence of such a dual telecommunications system is by itself proof—absent any market imperfections—of the need for such a dual system.

Network Utilization Principle 5: All services use the network in order to obtain the network surplus.

A network is used for only one reason: It is cheaper than any other alternative. Accordingly, the price charged to a firm by a network is influenced by the size and availability of the network surplus. The surplus is equal to the difference between what the self-provisioning of an alternative network would cost the firm—that is, its annualized incremental cost—minus the annualized incremental cost incurred by an existing commercially available network. The firm does not know the size of the network surplus as only the network owners or managers know its demand-adjusted incremental cost. All the firm knows for certain is its stand-alone cost and the announced price of the network.

The network, however, does know where its announced price to the firm is in relation to its incremental cost.[14] A cross-subsidy occurs when the price charged is below the network's incremental cost. Unless otherwise constrained by competition, market structure, or regulation, the network has the pricing flexibility to give a firm a price anywhere between its incremental cost floor and the stand-alone cost of the firm. The firm and the network have imperfect knowledge about each other's costs. A freely negotiated contract between the firm and the owners of the network indicates the comparative value of this surplus to both parties.[15] The surplus is one important source of the network's pricing flexibility.

As long as the network price stays within the preference range of the firm—which is between its understanding of the network's incremental cost and the annualized cost of self-provisioning—it will stay on the network. In a system in which all users of the system have perfect information about the cost of alternative networks, subnetworks, and nonnetworks for all other parties, the size of the network surplus received by each party will be known and agreed on within some margin of tolerance. In regulated public-switched telecommunications networks, government agencies or legislatures establish policies that authoritatively determine the amount of the network surplus applied to the prices charged to each customer. As regulated telecommunications become more competitive, these policies as well as the size and recipient of the surplus will change.

If every firm using a network had an identical usage pattern, reliability needs, and a similar location, then costs would be identical for each firm and, presumably, the prices charged would be identical. Because the use of a network is rarely exactly uniform by the customers of the network, costs incurred and prices charged also tend to vary by customer.[16] Armed with this knowledge, the network's actual and potential customers have a twin set of incentives. The first is to ensure that its firm pays the lowest possible price. The second is to ensure that the prices charged other customers do not cause the price charged to the firm to be higher than that expected to be achieved by negotiation. Discipline is enforced in any price-setting negotiations by Network Utilization Principles 1 and 2.

The agreed-upon price should be between the cost to the buyer of obtaining or constructing an alternative network and the cost to the owner of the network of providing the service. The difference between the cost-based prices charged that firm would have to assess itself to recover the cost of using or building an alternative network and those ultimately agreed to by the firm and the network owner are made possible by the network surplus. If no surplus is available, and the network owner posts prices that are above the prices charged by alternative networks, or the legitimate cost of the network is higher than available subnetwork or nonnetwork substitutes, then the potential customer should build or rent an alternative network or abandon its planned sale of network-based services. The network surplus is a natural by-product of and occurs in any decreasing-cost network. The surplus is available, in principle, to all customers and is the product of any economies of scale and scope that the network enjoys.[17] In regulated networks, government agencies determine the availability of the surplus.

The ability of a firm to obtain some or all of the network surplus is affected by a number of factors. These include, but are not limited to, the information available and the availability of substitutes. Both of these factors directly affect the bargaining power and ultimate outcome of negotiations. In a monopoly situation with no technologically equivalent substitutes available, the bargaining power of the firm is inferior to that of the network owner. In other more competitive market situations, the relative bargaining power of firm and network change accordingly.

It is important to recognize that the network surplus is not necessarily a cross-subsidy. Generally, a cross-subsidy is said to exist when one service has its price explicitly set above its incremental cost so that the price for another service can be set below its incremental cost.[18] In the case of a declining-cost, ubiquitous telecommunications network the highest possible price a network can charge a firm is equal to or below the firm's self-provisioning cost. Because of this fundamental feature, less incentive exists to charge prices below the network's incremental cost as the firm has already decided that the network is the least cost alternative. The only issue to the firm is how much of the network surplus it can obtain. A cross-subsidy in an unregulated network is only possible when significant information asymmetries exist or when the negotiating firms agree to provide a cross-subsidy when markets are imperfect.[19] The key pricing issue for the network owner is how to distribute the surplus in such a way that optimizes profits and customer retention.

Network Utilization Principle 6: A network is integrated and indivisible.

Unlike some other modes of industrial production, such as those found in retail stores and in the insurance and agriculture sectors, a network has as its core reason for existence the ability to uniformly, reliably, and automatically connect all of its customers with each other upon demand. Although some routes, services, or switches may be more heavily used than other routes, services, and switches, a pattern of differential network usage does not affect the fundamental connectivity and ubiquity that all networks sell. The need to achieve ubiquity for a self-defined area combined with the need to employ engineering optimization strategies in order to achieve cost advantages over self-supply options and alternative non-network-based competitors has the combined practical effect of making a network indivisible and integrated.

If an existing telecommunications network was broken into its constituent parts, the result would be either a set of point-to-point subnetworks or a set of nonoverlapping, but smaller networks. The smaller networks, however, would still be indivisible and integrated, but for a smaller geographical area. Once ubiquity is lost, a network loses its fundamental ability to connect all of its former customers. It is no longer integrated or indivisible. There is no economic-based presumption that subnetworks should become networks or that the disaggregation of existing networks into subnetworks is something to be avoided. Instead, the logic underlying the network utilization principles suggests that economic demand will be the initial basis for determining whether a firm needs a network, or a subnetwork, or a nonnetwork. It is the aggregate pattern of the demand for ubiquitous and nonubiquitous telecommunications services that determines the number, size, and mix of telecommunications networks and point-to-point subnetworks.

To examine this point further, it is important to ask if an optimally designed network that had the most efficient possible arrangement of lines and switching could be reliably disaggregated into its constituent parts for costing and pricing purposes. Imagine two central offices, A and B, linked to each other through a trunk connection. Could the customers of Central Office A convincingly argue that their costs are only the costs for Central Office A and not for the trunk line connecting the two offices? In this instance a trunk connection would not be built if sufficient demand from the customers of A and B did not exist. The linking of the two central offices indicates that the "local" ubiquity of the two previously stand-alone central offices has been replaced by the ubiquity represented by the combined A–B network. The size of the trunk connector simply represents the calling volume expected within specified technology, reliability, and quality-of-service parameters. It is the presence of the connecting trunk itself that indicates the demand to expand the ubiquity of the telecommunications network.

If only local ubiquity is desired by a critical number of firms, then networks consisting only of Central Offices A and B could be operated. To the extent, however, that the net increased demand for trunk A–B, tandem-based services provides revenues that allow a continued A–B network to price its "local" services and allocate its network surplus such that the prices for local are lower than stand-alone Central Office A or B, then the expanded network should prevail. To the extent that the increased ubiquity is not desired, or is not economically efficient, or does not produce lower prices, then the enlarged A–B network could fail.

This form of reasoning can be extended and generalized to deal with any network disaggregation proposal. If the decision to build facilities is irrational from an economic perspective, the customers will choose a least cost alternative that meets their communications needs. This includes decreasing their demand. What tends to make the network attractive to a firm is the traffic optimization capabilities of the network. A network achieves traffic optimization because of the "law of large numbers." Being able to take advantage of this "law" is one of the main reasons franchised utilities with an obligation to serve have been so successful. The large customer base of the network enables it to build proportionately less capacity because whereas every customer is connected to the network, not every customer has the same communications profile. Accordingly, a successful network can build a proportionately smaller network than might be built by a firm having a lower amount of traffic.

In order to further examine the indivisibility of a network, imagine the following instance. A firm desires that the ubiquitous network provide a service consisting only of a single unswitched line to a computer owned by another firm across the street. Could the stand-alone costs for this service be reliably identified and would these costs be a valid representation of the costs incurred by the ubiquitous network? These questions can be answered by using the network utilization principles identified previously.

A firm seeking to use the ubiquitous network's facilities, or to have the network construct facilities, seeks to use the network because the alternatives available to the firm are more costly. In particular, the network of a regulated telecommunications utility is able to provide the service at a lower price because it has valuable government-approved right-of-ways, along with repair, service, and maintenance facilities throughout its service territory. It has a sales, marketing, construction, and network operating ability that, within known error and reliability standards, is equally available to all customers within the area served by the network. The price charged to the firm seeking only an unswitched, dedicated line includes an appropriate share of what in aggregate it cost the network owners to attain the cost structure that enabled the network to be the least cost alternative. In an unregulated market the existing customers of the ubiquitous network will monitor the prices charged other customers in order to ensure that they are being charged a fair price and receiving their fair share of the network surplus. If the network owners decide to charge the firm requesting the single line a price less than its incremental cost, the existing and future customers of the network could have an incentive to bargain for the same price and to seek an alternative provider if the bargaining outcome is unsatisfactory.[20]

As noted earlier, one of the major reasons for the cost advantage enjoyed by a network over alternative means of providing the same service functionality lies in the traffic-routing capabilities of the network.[21]

3. APPLICATION OF NETWORK UTILIZATION PRINCIPLES TO THE PROVISIONING OF RELIABILITY FOR TELECOMMUNICATIONS NETWORKS

3.1 The Significance of Reliability in Telecommunications Networks

Reliability is an intrinsic feature of network and point-to-point subnetworks. State and federal regulators and international standard-setting bodies have traditionally been concerned with ensuring the reliability of the public-switched telecommunications network.[22] Regulators have established quality-of-service standards that specify the reliability minimums for a network. Typical reliability standards focus on items such as dial-tone delay, call completions, directory assistance, and interofficial transmission.

Regulators and network users have worried that the cost-control incentive of price caps and other regulatory pricing reforms may cause a network operator to invest less in reliability-increasing investments (such as might be needed for new and expensive generations of error-monitoring software) and to spend less on maintenance. Network owners have generally responded to these expressed concerns by stating that the reliability of the network is above existing standards and that this level of reliability should increase in the future. More importantly, network owners note that reliability is so inextricably related to their ability to sell ubiquitous service that it would be economically irrational for them to allow network reliability to degrade.

In order to meet the twin challenges of the information age and the emergence of competition, local exchange companies have made massive and sustained modernization investments. Modernization is generally thought to increase the reliability of the public-switched network. One unintended but inescapable consequence of the modernization is the increase in the amount and concentration of traffic on certain links or digital switches. Concentration of traffic may be the single most important way to increase the economic and engineering efficiency of a network.

Unfortunately this higher level of traffic concentration can increase the magnitude caused by a disruption, even with an increase in reliability levels. Fiber lines are economically attractive because many more calls can be handled on one fiber strand than can be done for the same amount of copper wire. A single fault or disruption, accordingly, can interrupt many more calls than may have been possible previously. The service outages in 1991–1992 for U.S. carriers— with AT&T's massive New York outage receiving the most attention—illustrates the magnitude and impact of service disruptions possible from a relatively small number of faults.

Following these outages, the Federal Communications Commission (FCC), responding to Congressional concerns, established a Network Reliability Council (NRC) in December 1991 to provide the FCC with expert advice and recommendations. One of the recommendations from the NRC was to establish a new threshold for reporting outages when 30,000 lines are affected and to refer reliability issues to appropriate industry forums for further analysis and recommendations.[23] The FCC has also ordered in its price caps proceedings certain modifications to its service quality and infrastructure reporting.[24] Concern about network reliability has also resulted in governmental action in the United Kingdom and Japan. For example, Gupta reported that in Japan tax and depreciation incentives have been provided by legislation to increase investments that improve network reliability.[25]

Insufficient attention, however, has been paid to improving the understanding of the cost structure underlying the provisioning of various levels of network reliability. Nor has sufficient attention been given to the impact of the modernization decision rules on network reliability. Both of these issues are addressed later through the application of the network utilization principles identified previously. Use of these principles allows regulators, utilities, and firms to make more efficient reliability pricing decisions. Current pricing practices may send the wrong price signal to regulators, utilities, and firms and may result in the inefficient provisioning and pricing of reliability. As telecommunications networks become increasingly complex, competitive, and technologically advanced, the successful resolution of reliability issues becomes even more important.

3.2 Average and Special-Purpose Reliability

Reliability is an important and intrinsic feature of telecommunications networks. Firms and individuals choose to use networks because of the networks' ability to reliably communicate with all other customers of the network. Networks do not, of course, have the ability to provide service that is always perfectly reliable. Even so-called self-healing, ring-based networks do not have the ability to ensure perfect reliability, but they do so at the added cost of essentially duplicating facilities in a way that provides rapid rerouting of interrupted telecommunications traffic.[26]

If a network's customers all had the identical need for reliability, no particularly compelling reason would exist for identifying a separate cost or price for reliability. When the need for reliability is uniform, stable, and known, its pricing will mirror the network pricing parameters described earlier. As long as the price charged for the use of a network service with a given degree of reliability is less than that available by self-provisioning, or from network and nonnetwork competitors, no new issues arise.

Because of the very wide array of services available from telecommunications networks, the different construction and operational costs incurred for different levels of reliability, and because of the different value firms place on each service, the pricing of reliability has become an important issue. From an engineering perspective, two kinds of reliability are possible: average system reliability (ASR), and special-purpose reliability (SPR). ASR is the average ability of any part of the network to deliver uninterrupted communications upon demand to and from any part of the network. The failure or error rate is known and randomly distributed for certain types of facilities. In general, all customers of the network have the same average level of service reliability.

Special-purpose reliability is quite different. By way of analogy, ASR is to average cost pricing as SPR is to incremental cost pricing. SPR occurs when, say, a fiber-token ring is constructed to serve specific, generally large-volume, customers. The ring typology is an efficient way to provide alternative routing or switching so that traffic that otherwise would have been interrupted can reach its intended destination. This kind of reliability is special because only a portion of the network is generally served. If the whole network had a token-ring structure, then this special reliability would become ASR. Furthermore, although all networks have ASR, they do not necessarily have SPR.[27]

For regulated and unregulated networks, a tension exists between the network customers that desire only ASR versus those who need SPR. In regulated, public-switched telecommunications networks this largely parallels the debate over plain old telephone service (POTS) and enhanced telecommunications services. In a regulated network facing little competition the network customers that require higher or special levels of reliability would be in favor of pricing policies that spread the cost of special reliability across all network customers. Here the logic is that special reliability is a network feature that could eventually be available to all network customers and that an intertemporal shift and socialization of costs is in the public's interest and in the eventual interest of all network users. POTS customers view SPR as a premium service that is a private good and should be strictly paid for by the direct users of SPR. Furthermore, they see this type of averaging as running counter to the cost-causation and unbundling principles that are necessary to make a network efficient. POTS customers see no necessary externalities in efforts to achieve SPR and desire only to pay for ASR. Disputes over the apportionment of the costs of average and special purpose reliability are resolved by commission rulings where networks are regulated.

Unregulated telecommunications networks in competitive environments face additional constraints. Presumably, average levels of reliability are known and are among the factors used by a firm in selecting a particular network. If the choice is between one network with 100% token-ring backup versus a network with no token-ring backup, then the firm can sort out its price and reliability preferences and select the network having the best combination. In hybrid networks in which only selected customers have token fiber-ring backup the choice is more complicated but follows the same price/reliability, decision-making logic. The key reliability pricing issue lies in determining why, if a non-token-ring network is equally available, a firm that is not directly serviced by the token ring chooses the particular token-ring network. As long as the SPR customers pay for their special service, no particular problem occurs. Indeed, for a hybrid network, it should normally be expected that the SPR self-provisioning costs are higher than those firms needing only ASR.

If SPR is being thought of and treated as ASR and is being phased in and the lead time is acceptable to a non-token-ring using firm, then a price/reliability balance may be achieved. It would be an unstable and unsustainable condition if an unregulated network charged a portion of the SPR costs to those who only demand ASR. With competitive options available, the ASR customers would seek a network whose prices better reflect the reliability levels actually desired and used. Only if no other networks were available, or if the cost of a self-supply option was unacceptable, or if the incumbent network had significant market power, or if the cost of a self-supply option was unacceptable, would an unregulated network be able to enforce this type of pricing. The network owner has ASR/SPR pricing freedom when it allocates the network surplus as demonstrated through the previous examination of network utilization principles.

3.5 Modernization Decision Rule

Reliability is inextricably intertwined with the type of network technology. Fiber, radio-wave, and digital-switching technologies offer increased reliability over copper lines and older switching technology. To the extent that increased reliability lowers network costs, the main issue is how the network owners use the reliability-driven cost savings, which will depend on the goals of the network owners and the competitiveness of their markets. A larger problem occurs if increasing reliability—either for average or special-purpose reliability—also increases a network's costs.

In standard engineering and economics texts, modernization decisions are uniformly described and based on clear, elegant, and powerful decision rules that say that a modernization investment should be made if and only if it will increase net, future-revenue streams. Following this modernization decision rule, a network will not invest in a new technology—one having a known reliability cost structure—unless doing so will profitably increase future revenue streams. Unless the modernization analysis is flawed or conditions change (say, the forecasted demand does not occur), both average system and SPR costs and resulting prices should be expected to be more favorable than those existing before the modernization investment. Otherwise, as indicated by the network utilization principles, those firms disadvantaged could go to an alternative network. Accordingly, if the modernization decision rule is followed, only forecasting or data problems can cause problems with either average systems or SPR costs. Assuming the availability of competitive networks, if the modernization rule is not followed, then the network may lose customers and may fail.

In Fig. 6.1, possible outcomes for networks correctly following the modernization decision rule in competitive and noncompetitive markets are displayed. Network failure (depending on the magnitude of the modernization investment) occurs most quickly in competitive markets when unsustainable and uneconomic modernization decisions are made. Because a network's customers have options, they can easily migrate to networks that have followed the modernization decision rule correctly.

A stable situation occurs when no competitive options are available. Unless self-supply is a viable option or new entrants are eventually allowed or induced in, the monopoly or otherwise noncompetitive network does not experience customer loss. It may, however, experience a decrease in demand. It is the elasticity of the demand that will determine the impact of inefficient modernization prices.

The most successful outcome occurs when competitive markets exist and the modernization decision rule is followed. Here both the network and the network's customers are better off. Because there are legitimate options readily available, application of the network utilization principles indicates that the network owners

	Did the Network Use the Modernization Decision Rule for New Technology?	
Are Competitive Networks Available?	Yes	No
Yes	**Successful** outcomes for network and firm because all networks will adopt new technology and prices. Pattern of savings distribution is known.	**Failure** and predicted migration to networks correctly following modernization decision rules in order to obtain more favorable prices.
No	**Successful** outcome, but only incumbent network is available and pattern of distribution of savings from modernization unknown.	**Stable** situation as no migration possible. New entrants, or self-supply options will be considered unless prohibited by regulators or precluded by incumbent network's market power.

FIG. 6.1. Possible outcomes in competitive and noncompetitive markets for alternative applications of modernization decision rules.

in this instance do a more efficient job of distributing the increase in the network surplus due to the modernization investment. The competitive nature of the market provides the necessary incentives for this successful outcome. A less successful outcome also occurs when markets are not competitive but the modernization decision rule is followed. It is the unknown distribution of the modernization savings that makes this example problematic and not as desirable.

4. CONCLUSION

In practical terms, as long as a network is less costly than reasonably available, self-supply options, and if no real competition exists, network owners have considerable freedom in choosing their pricing strategies for any service that they sell, including reliability.

Deciding who pays for reliability is determined first by the network utilization principles, in which a firm determines whether self-supply or the use of an externally provided telecommunication network is in its best interest economically. The second factor is the availability of competitive networks. Reliability likely will be priced in ways that reflect the market for telecommunications networks. The traditional postal telephone and telegraph agencies, for example, faced no significant competition and priced their services, including reliability, largely by administrative fiat. It is now widely thought that long-distance toll networks are competitive enough so that competitive pricing strategies (including those needed for reliability) are followed and prevalent.

Two other factors affecting the pricing of reliability are the adherence to modernization decision making rules and the different levels of reliability that may be present in a network. Ideally the cost of reliability should increase only when modernization rules are not followed. The more difficult issue is paying for average-system reliability versus special-purpose reliability. As long as average reliability is cheaper than self-supply and equivalent to other networks, no special pricing problem exists. When no competitive networks are available and self supply is not a viable option, the network owners have considerable pricing freedom when providing services requiring special-purpose reliability. Only when alternative networks exist do network owners have to price reliability competitively. The basic network utilization principles can be restated in the following six reliability pricing rules:

1. Determine whether a network or subnetwork is needed.
2. Determine if self-supply or the use of an externally provided telecommunications network is the least costly alternative available to the firm.
3. Determine the availability of competitive network.
4. Determine whether the networks available to the firm follow the modernization decision rule.

5. Determine whether the firm requires average reliability or special-purpose reliability from the network for the network service it wishes to purchase.

6. Choose the pricing option that offers the lowest cost reliability for the firm among the self-provisioning, network, subnetwork, and nonnetwork options.

From a public interest perspective, policies that encourage self-supply options or the emergence of competitive networks are important for pricing reliability efficiently. The recent opening up of the Class 5 Office bottleneck to facilities-based competition is extremely important in this regard. Self-supply options should be increased in open network architecture-type, unbundling approaches. As competition strengthens, these and other similar approaches should make existing and emerging networks more efficient and result in a better pairing of reliability costs and prices.

The national information highway will use a number of networks and sub-networks, each having different cost and reliability parameters. Common standards can help with establishing minimum reliability floors, interconnection rules, and interoperability protocols. The network utilization principles and reliability decision rules developed here can help regulators, utility managers, and telecommunications providers develop prices that track the cost of providing the increased reliability likely to be required by the users of a national information highway.

ACKNOWLEDGMENTS

This chapter was prepared under an Ameritech Fellow award made by The Ohio State University Graduate School. The views and opinions expressed herein are solely those of the author and are not necessarily those of The National Regulatory Research Institute, Ameritech, or The Ohio State University.

ENDNOTES

1. Jean-Michel Guldmann, "Modeling Residential and Business Telecommunications Flow: A Regional Point-to-Point Approach," *Geographical Analysis* 24 no. 2 (April 1992): 121–141.

2. In addition to the assumption of an economically rational firm, this analysis is based on two other important assumptions. The first is that the networks described are unregulated networks and that competitive networks are available, as are substitute delivery mechanisms. The second assumption is that there is a uniform availability and usage of the same telecommunications technology and industrial organizational structure.

3. Peter W. Huber, *The Geodesic Network* (Washington, DC: Antitrust Division, U.S. Department of Justice, 1987).

4. A company does not necessarily have to serve every possible customer, but rather must have the facilities-based ability to link all firms desiring to be customers in a given geographic area upon request.

5. A business located just outside the railroad company property must still be served by another telephone network. If the unit of comparison is limited to railroad facilities, then this would constitute a network because it is ubiquitous. If the unit of analysis is a region, state, or nation, the railroad instance could only be considered to be a subnetwork. A point-to-point subnetwork is also economically efficient, otherwise an economically rational actor would not have built the system and would have found it more efficient to continue to use the ubiquitous network.

6. The existence of a viable network does not mean that the network is either a natural or franchised monopoly, only that it provides ubiquitous service. A test of subadditivity to determine whether single provisioning is less costly than that by two or more providers is unnecessary here because of the sequence of decisions made by the firm seeking to buy telecommunications services. By the time network utilization principle 3 is relevant, the firm has already made decisions about its need to buy ubiquitous services and its "build versus buy" decision such that subadditivity information is irrelevant. Furthermore, subadditivity is generally not as useful in markets with competitive networks.

7. Markets with competing networks can be stable or experience successes and failures even when using the same technology and type of industrial organization, design, or management.

8. Lines can be wire-based, coaxial, fiber, or wireless radio options such as cellular, personal communications systems, or other radio spectrum-based options.

9. R. F. Rey, ed., *Engineering and Operations in the Bell System* (Murray Hill, NJ: Bell Laboratories, 1983).

10. Harris identified the scale, scope, network, and learning economies that occur in the public-switched network that give it significant cost and engineering advantages over alternative communications modes. These features and advantages exist in equilibrium in competitive, transitional, and monopoly markets. These advantages apply whether a firm is selecting one telecommunications service or the entire range of telecommunications service desired by the firm. Robert G. Harris, "Telecommunications Services as a Strategic Industry," in Michael A. Crew, ed., *Competition and the Regulation of Utilities* (New York: Kluver Academic Publishers, 1991).

11. Diseconomies occur because the cost of low-traffic channels on per-call basis is greater than the average per-call cost of all calls on the network.

12. The analogy to be drawn here is that the cost structures that produce the economic efficiency of the network are the centripetal forces that keep the network whole and cohesive. The diseconomies present in portions of the network that do not have uniform costs and traffic flows are the centrifugal forces that can cause the network to be reconstituted. Absent any other constraints or goals, an economically rational firm weighs the net of these centrifugal and centripetal forces and either stays on the network or participates in some form of a subnetwork or nonnetwork mode. A network with ubiquitous service has a different economic structure than a subnetwork. Subnetworks tend to be more uniform, being composed, for example, of high-volume routes only.

13. By various contractual means subnetworks may be linked with networks or to other subnetworks. In a system in which prices follow costs the subnetwork will pay its fair share of network cost. Recall, however, that the continued existence of a subnetwork proves it is economically viable for whatever specialized purpose it is put to by its owners and customers. It is only when the subnetwork desires additional or ubiquitous services that it seeks some form of network access.

14. Gerald R. Faulhaber, "Cross-Subsidization: Pricing in Public Enterprises," *American Economic Review* 65 no. 5 (1975): 966–977.

15. Network owners can engage in a number of pricing strategies to deal with contestable and other types of markets. See William J. Baumol, John C. Panzar, and Robert D. Willig, *Contestable Markets and the Theory of Industry Structure* (New York: Harcourt, Brace, Jovanovich, Inc.,

1982). The value placed on the surplus price agreed to by the owner of the network is the end result of an optimization process that weighs and evaluates the preferences and risk-taking style of the network owner and the firm. It also involves judgments about the elasticity of demand for the services provided by the network and an appraisal of the availability of substitutes.

16. Recall that it is assumed that prices charged are cost based. Without this assumption a network may choose to offer, or be ordered to offer by a governmental agency, a price to a customer that does not necessarily recover the cost of service to that customer. Lifeline rates and rates charged to the hearing-impaired are examples of rates that reportedly do not cover the cost of providing the service. In regulated networks higher prices are charged, for example, on the non-hearing-impaired services sold to recover any loss. In unregulated systems the owner of the network may pursue a variety of strategies to cover any such self-imposed losses.

17. It is difficult to know exactly what the size and scope of the economies enjoyed by a network are. To do so would require the accurate and complete costing of a nearly infinite variety of alternative networks and point-to-point subnetworks. For all practical purposes it seems sufficient to assume that the continued economic viability of an unregulated network is enough proof that a network has positive but unknown economies.

18. Faulhaber, "Cross-Subsidization"; Sanford V. Berg and Dennis L. Weisman, "A Guide to Cross-Subsidization and Price Predation: Ten Myths," Paper (November 7, 1991).

19. It does not matter what costing method the firm or network employs as long as it believes the information to be sufficiently accurate and its subsequent network utilization decisions are based on this information.

20. Different network customers will have different perceptions about the attribution and assignment of costs and prices and will bargain accordingly. If a firm or group of firms possesses market power, it may be able to obtain network services at a lower price than would otherwise be the case without it having market power. It is here that the existence and allocation of the network surplus becomes important. As long as a firm pays less than its stand-alone, self-provisioning cost and more than its perceived incremental cost, the only issue of note is how much of the network surplus will be applied against the prices charged to the firm.

21. If a network is unable to offer a lower cost-based price to a firm than another network or point-to-point subnetwork, it is, by definition, not the least cost provider. This distinction is important as ubiquitous networks often compete with each other and with specialized subnetworks.

22. Bruce Armstrong, US WEST Service Quality, Handout at the Regional Oversight Committee, Minneapolis, MN, September 1993.

23. For example, during the first quarter of 1993, some 431 outages affecting more than 30,000 lines were reported by the NRC. This compares to outages for the first quarter of 1992 and an average of 44.9 outages per quarter from January 1992 to June 1993. United States Telephone Association, President's Report (Washington, DC: United States Telephone Association, August 27, 1993).

24. Federal Communications Commission, Modifications to Service Quality/Infrastructure Reporting (Washington, DC: Federal Communications Commission, July 7, 1992).

25. Suren Gupta, Japan Telescene, Newsletter (Tokyo, Japan: InfoCom Research, Inc., 1993).

26. Token rings differ from the traditional star and bus network typologies normally used by public-switched telecommunications networks and cable television distribution networks. Token rings connect users via distribution panels and a line configuration that establishes a ring such that any fault or path interruption is automatically detected and traffic is instantly rerouted through a bypass switch. Operator intervention is not required because the decision to bypass and follow an alternative path is made by specific failure-detection circuits. See John D. Spragins, "Telecommunication Network Reliability Models Based on Network Structures," unpublished paper (Clemson, SC: Clemson University, 1992), p. 23; and Yael J. Assows and Vikram R. Saksena, "Economic Analysis Architectures in Two Tier Data Networks," IEEE Network (May 1989): 13–21.

27. The special-purpose reliability provided by a self-healing token ring is designed to produce 99.999% reliability. Richard Tomlinson, "Impact of Competition on Network Quality," Presentation at the Quality Reliability of Telecommunications Infrastructure Conference of the Columbia Institute of Tele-Information, Columbia University, Graduate School of Business, April 23, 1993.

REFERENCES

Armstrong, Bruce, *US WEST Service Quality*. Handout at the Regional Oversight Committee Meeting, Minneapolis, MN, September 1993.

Assows, Yael J. and Vikram R. Saksena. "Economic Analysis of Robust Access Network Architectures In Two Tier Data Networks." *IEEE Network* (May 1989): 13–21.

Baumol, William J., John C. Panzar, and Robert D. Willig. *Contestable Markets and the Theory of Industry Structure*. New York: Harcourt, Brace, Jovanovich, Inc., 1982.

Berg, Sanford V. and Dennis L. Weisman. "A Guide to Cross-Subsidization and Price Predation: Ten Myths." Paper. November 7, 1991.

Faulhaber, Gerald R. "Cross-Subsidization: Pricing in Public Enterprises." *American Economic Review* 65 no. 5 (1975): 966–977.

Federal Communications Commission. *Modifications to Service Quality/Infrastructure Reporting*. Washington, DC: Federal Communications Commission, July 7, 1992.

Guldmann, Jean-Michel. "Modeling Residential and Business Telecommunications Flows: A Regional Point-to-Point Approach." *Geographical Analysis* 24 no. 2 (April 1992): 121–141.

Gupta, Suren. *Japan Telescene*. Newsletter. Tokyo, Japan: InfoCom Research, Inc., 1993.

Harris, Robert G. "Telecommunications Services as a Strategic Industry." *Competition and the Regulation of Utilities*. Michael A. Crew, Editor. New York: Kluver Academic Publishers, 1991.

Huber, Peter W. *The Geodesic Network*. Washington, DC: Antitrust Division, U.S. Department of Justice, 1987.

Rey, R. F. *Engineering and Operations in the Bell System*. Murray Hill, NJ: Bell Laboratories, 1983.

Spragins, John D. "Telecommunications Network Reliability Models Based on Network Structures." Unpublished paper. Clemson, SC: Clemson University, 1992.

Tomlinson, Richard. "Impact of Competition on Network Quality." Presentation at the Quality Reliability of Telecommunications Infrastructure Conference of the Columbia Institute of Tele-Information. New York, NY: Columbia University Graduate School of Business, April 23, 1993.

United States Telephone Association. *President's Report*. Washington, DC: United States Telephone Association, August 27, 1993.

Chapter 7

Rate Regulation and the Quality of Cable Television

Thomas W. Hazlett
University of California, Davis

1. INTRODUCTION

Cable television price controls have had pronounced effects on the way cable systems deliver their video signals. This has been neatly demonstrated over the past decade or more in a series of experiments that policymakers have fortuitously granted: deregulation in the 1970s and 1980s, reregulation in the 1990s. In each instance, service suppliers have rationally sought to adjust their service menus and even technical delivery mechanisms in rather dramatic fashion. For the most part, quality changes induced by regulatory shifts produced entirely predictable results, but there have been some counterintuitive surprises. In any event, the changing regulatory environment has given researchers ample opportunity to examine the ways in which market forces operate under exogenous pricing constraints.

This chapter examines this subject matter by first examining the theory of price controls. Then it outlines three instances in which a regime switch occurred in the cable television marketplace: a 1979 deregulation of California cable television rates; the 1984 Cable Act, which deregulated prices nationally; and the 1992 Cable Act, which reregulated cable rates nationally. In each instance the impact of rate controls on consumers is examined, with emphasis placed on the important role of product quality.

2. THE THEORY OF RATE REGULATION APPLIED
TO CABLE

When binding price controls are levied on a good or service, demanders and suppliers react—at a general level—in fairly standard ways.[1] Demanders are willing to bid up prices above rate-controlled levels and express this willingness with payments that are officially "nonprice" offers. Such offers may entail both transfers (attempts to pay suppliers in either legal or illegal means, effectively raising the price paid above the controlled level) and waste (such as the opportunity costs expended by consumers queueing in shortages). Both sorts of expenditures may be thought of as rent seeking, but only the latter entails social inefficiency. The first sort may be thought of as a simple (low-cost) evasion of the price-control mandate.

Suppliers also react. As binding price controls will increase the quantity demanded, ceteris paribus, the seller is tempted to withdraw costly inputs, and thus regain the pre-control equilibrium. In this, the supplier is constrained by the cost of input substitution (i.e., factor mobility). In the limiting case assuming zero transaction costs, a supplier facing regulated price of γP_0 (where P_0 = uncontrolled price and $\gamma < 1$) simply reduces the size of the package sold by $(1 - \gamma)$. Thus, price controls are rendered moot via quality depreciation. This supplier reaction holds both in the case where supply is upward sloping (and binding controls produce excess demand) and in markets where $P > MC$ for the last unit sold. In general, the profit-maximizing firm, under binding price controls, is led to engage in such quality-lowering substitution until the marginal gain (from total costs declining) equals its marginal cost (from lower revenues as consumers decrease their demand prices).

This straightforward but powerful result suggests that, even abstracting from consumer efforts to bid around legal constraints, price controls will never be binding so long as suppliers' inputs are perfectly mobile.[2] The operational corollary for policymakers is that price controls cannot be successfully implemented without some practical and/or imposed constraint on product quality. The package of services on which the price control is levied must be "sticky" with respect to quality, otherwise the "controlled" product will simply metamorphose into a new, deimproved, "decontrolled" product. Where resources are mobile but imperfectly so, the same incentive for suppliers to depreciate quality will obtain, yet the transition will entail transaction costs.

The classic illustration of this problem is offered by rent controls in residential housing markets. Policymakers have long known that municipalities instituting rent controls must undertake to monitor a host of second-order adjustment margins. Those regulations circumscribing nonprice (or extra lease) bidding for apartments (e.g., key money, bribery, discrimination) have often been honored mostly in the breach (Hazlett, 1985). However, rules attempting to control product quality have been at least partially effective. Most notable are laws limiting the ability of landlords to withdraw rental units from the market altogether. Condo (or co-op) conversion laws are notably strict in communities with stringent rent controls. Also important, however, are arbitration mechanisms that seek to monitor landlord behavior regarding "discretionary" payments for maintenance.

The apartment market has at least one very large difference when compared to the cable television market: the relative fixity of apartment services. Once the structural investment is sunk, the housing services flowing from a building cannot be instantly depreciated by the curtailment of maintenance expenditures.[3] Cable television services, however, entail both the transport function over cables (sunk similar to housing structures) and the provision of video programming. Those service menus are comprised of inputs that are highly mobile.

Not only is quality difficult to measure[4] and, hence, monitor, cable regulators are legally constrained from exercising any discretion over programming quality due to both federal regulations and the U.S. Constitution.[5] Although public regulation has been fraught with difficulty (insofar as producing a price lower than the unregulated alternative is concerned) even when resulting from an intensive effort to determine input costs and optimal production patterns (see Stigler & Friedland, 1962), setting price without any ability to control the unit being sold appears a Herculean task.[6]

Another potentially important aspect of cable rate regulation is institutional. Quite distinct from utility rate regulation, cable television systems have not been regulated on a strict rate-of-return basis. Instead, franchise agreements (generally controlled by municipalities but in some cases involving state regulatory authorities[7]) have required that the franchisee receive permission to raise rates. Typically, cable rate regulation involves a city council voting simply to approve or disapprove price increases, and it does not involve elaborate fact finding with respect to a system's underlying cost base. Even the 1992 federal reregulation (which mandates the Federal Communications Commission to create rules, or guidelines, that are enforced by localities) sets rate caps rather than rate-of-return formulae.[8] The institutional structure is fundamentally different, then, from the cost-plus environment created by rate-of-return regulation, which has given rise to the overinvestment incentives described as the Averch–Johnson effect (Averch & Johnson, 1962).

The Cable Television Package

Cable program services are delivered in three broad groupings: basic services, premium channels, and pay-per-view. The most popular basic package contained an average (nationwide) of 35 channels in the General Accounting Office's 1991 survey of cable rates.[9] These include the locally available off-air (broadcast) signals (such as ABC, NBC, CBS, Fox, and PBS, and the independents), as well as cable-only networks (such as CNN, ESPN, Discovery, A&E, USA, and

MTV[10]). This basic package is the product that has been regulated or deregulated as dictated by regulatory policy. Premium services (such as HBO, Showtime, or Disney) are delivered on an á la carte basis and have been explicitly exempt from price regulation since the late 1970s.[11] Pay-per-view falls under this latter exemption.

Basic Tiers

Within the basic package there are varying tiers of service. The *limited basic* service has characteristically included a bare-bones menu of off-air broadcast signals and local origination/public-access programming (sometimes including C-Span). This level of service has constituted a small portion of the market, although some have placed subscribership as high as 6%–9% of all cable households.[12] *Basic* service has often included a much larger complement of cable-only program networks. *Expanded basic* has often included an additional tier of relatively expensive basic programming, including regional sports networks. These latter two categories are generally what people refer to when they talk about cable television service.

Retransmission and Customer Service

In terms of identifying product quality, two final components of the cable television package are important. The first is the quality of broadcast signal retransmission. Because the cable household will spend at least half of its viewing hours watching broadcast signals over cable, it may be assumed that the increased signal quality of off-air channels received over wireline systems will account for some of the service for which consumers pay. (This is also implied by the fact that cable penetration rates are generally highest in areas in which off-air signals are not easily receivable via roof-top antennae.) The second is that cable companies can raise or lower the quality of their service. They can provide faster, more reliable technician appointments, have more or fewer signal interruptions, answer their customer service telephone lines faster or slower. Both elements of product quality are real, yet exceptionally difficult to monitor objectively. Hence, they will not be an explicit part of the analysis conducted herein.[13] The implicit assumption is that quality changes in measurable dimensions of product quality are highly correlated with quality changes in unmeasurable elements. When means of accounting for these aspects of quality are available, then the analysis may be appropriately expanded.

The response to price controls by demanders will be obviated if (a) suppliers can elastically respond to controls with quality changes that eliminate excess demand, and/or (b) the sector exhibits market power (with prices set above marginal costs) such that suppliers are willing to sell additional output at lower prices if constrained via price controls. It is apparent that at least one of these factors is operative in cable markets, as the period of rate controls has not witnessed excess demand: All the consumers willing to pay the controlled rate have been able to receive service. Hence, it has not been necessary for consumers to engage in "nonprice bidding" behavior.

The supply side is more interesting, for cable systems have been freely able to adjust their products in response to price controls. First, cable programming products are highly mobile resources. As noted earlier, the system sending television-viewing products has sunk little investment into its software and may change its program menu at low transactions cost. This menu may be changed by retiering, adding to or subtracting from the total number of networks offered, changing the quality of programming on given networks, or some combination of these factors. The predictable implication of price controls include the following:

1. When binding rate controls are imposed: systems will shift programming to unregulated tiers or á la carte status, or systems will reduce program expenditures and offer lower quality program services on regulated tiers.

2. When rate controls are eliminated: systems will shift programming to collapse tiers and eliminate á la carte services, or systems will increase program expenditures and offer higher quality program services on (formerly regulated) basic tiers.

The efficiency argument for price controls must rely on the existence of monopoly power. (In competitive markets, the argument for price controls must rely on distributional issues, as total welfare will predictably decrease.) That is, forcibly constraining maximum prices charged by a supplier that otherwise sets prices above marginal cost[14] will increase output (and increase the sum of producers' and consumers' surplus) ceteris paribus. Yet, it is understood that the demand curve may well shift downward as the supplier reacts to controls by withdrawing inputs. The interesting question for welfare analysis is whether demand shifts are sufficiently small so that the net effect of the price controls is that the quantity of output sold increases. Only if price controls increase output will one be confident in concluding that regulation lowered effective (quality-adjusted) prices. Hence, I now examine the evidence to discern the ways in which price-controlled cable companies modified cable television service in response to changing regulations and to observe what overall impact on equilibrium quantities such regime switches appear to have had.

3. THE CALIFORNIA DEREGULATION (1979)[15]

A 1979 California statute unleashed cable subscriber rates from the control of local franchising agreements at the discretion of the operator. In AB699, the California Cable Television Association procured freedom to price in a compromise with public-access lobbyists. In exchange for an annual fee per subscriber per year, any cable system within the state could elect to free itself from existing municipal rate controls (although systems with very high penetration[16] levels were then supposed to raise rates no higher than the statewide average rate). The money paid went to a foundation to subsidize local origination/public-access programming on cable systems throughout the state.

Over the 1980–1985 period, only about one-fourth of California's cable systems elected to pay the modest fee of 50 cents per subscriber per year to deregulate. The California State Legislature appointed two official study groups to analyze the episode, and both concluded that the price increases emanating from deregulation were "innocuous." Ironically, the studies nonetheless dubbed the reform a failure because too few systems had deemed regulation sufficiently onerous to pay to escape it. (Hence, a disappointing program subsidy fund was created.) Moreover, the availability of the deregulation safety valve had made it more difficult for local governments to enforce cable franchise terms. Greater reneging on agreements by the cable firms was cited as the chief economic outcome.

The Data Supporting These Findings

The mean annual price increase for systems that elected to become deregulated minus the same-year, mean annual increase of nonderegulated systems, weighted by system subscriber size and summed across all 6 years, was 10.22% (statistically significant at the 99% level). In that the deregulated sample is likely to be biased (firms crossing over to deregulation are predictably those systems with the most to gain from freedom to price unconstrained), there appear modest price changes associated with the deregulated sample. More interesting, perhaps, is the fact that output, as measured by penetration, increased in the deregulated systems (by a statistically insignificant amount) relative to same-year, nonderegulated systems.[17] As a regime switch from effective regulation to free-market pricing would be accompanied by restriction of output, these results are highly suggestive.

What they imply is that although prices increased in the newly deregulated systems, quality of service was increasing *pari passu*.[18] That penetration did not fall in systems opting out of price controls, even over a 2-year period, implies that regulation was not effectively constraining prices. If it had, output restriction would have accompanied the price increases visibly attendant to deregulation. Instead, the price-control mechanism appears to have constrained service/quality elements of the cable television package. When controls were removed, quality improved to offset price increases, and output stayed the same or increased (by a statistically insignificant amount).[19]

This is actually the analytical rendition of the public arguments made on behalf of AB699 in 1979. The legislature was told that local government rate regulation was creating a drag on cable system capacity upgrades and delaying the offering of the numerous new satellite networks then becoming widely available. As seen in Table 7.1, the number of such national program channels tripled between 1978 and 1980. On the operative quality margin, cable operators faced a choice between depreciating existing capital versus expanding channel capacity and delivering many more signals. The existence of local rate regulation added a regulatory tax on the decision to expand. Even when rate increases were routinely granted, the mere process of having to ask permission could impose significant costs. These might include compliance with demands to subsidize local origination/public-access services and to provide other nonremunerative amenities. (Often these were provided for in the cable franchise, but were actually supplied according to the relative importance of placating local officials.) Legal representation, litigation expense, and campaign contributions may also have been part of the expected cost of rate-increase requests. Moreover, the process imposed uncertainty on long-term financial decision making, thereby increasing business risk.

It is clear that in a dynamic industry rate controls may retard investment and growth, as firms evade the impact of controls by economizing on new technologies. Here, regulatory oversight would be severely crippled because the restriction of output is never observed. Instead, the carrier responds by failing to make efficient capital investments. This is analogous to a reverse Averch–Johnson effect. Whereas under rate-of-return regulation the firm overinvests so as to

TABLE 7.1
Cable Network Growth

Year	Basic Nets (NTIA)	Total Nets (NTIA)	National Video Nets (NCTA)
1976			4
1977			5
1978			9
1979			18
1980			27
1981	25	34	38
1982	35	46	41
1983	27	37	41
1984	26	35	49
1985	32	42	50
1986	44	55	60
1987	52	65	73
1988			73
1989			73
1990			74
1991			78
1992			78

Source: NTIA 1988, p. 11; National Cable Television Association [NCTA], Cable Television Developments (June 1993), p. 7-A. The NTIA and NCTA count national video networks differently, as seen. The total network count, in either case, includes basic and premium channels. The NCTA count includes superstations.

obtain a guaranteed profit on a larger capital base, under price caps with ineffective quality monitoring the firm underinvests. The regulated firm thus stints on capital expense, raising profits.

4. THE FEDERAL DEREGULATION (1984)

Federal rate deregulation began with the Cable Act of 1984, which achieved two major cable industry goals: It preempted local rate controls, and it strengthened incumbents' rights to franchise renewals by placing the burden of proof (for nonrenewal) on municipalities subject to procedural safeguards for franchisees.[20] In practice, this made it virtually impossible for a cable operator to lose his or her license; moreover, it significantly lowered the cost of rent-seeking (or rent-protecting) actions for the franchisee by decidedly shifting the legal burden.

The 1984 Cable Act has commonly been portrayed as a great triumph for incumbent cable operators (see, e.g., Powe, 1987).[21] That conventional wisdom is not challenged here. But the question remains: What was the source of the industry's victory? Although the question appears trivial by most journalistic and political accounts, which merely posit that the decontrol of prices has resulted in a textbook case of prices rising from competitive to monopolistic levels,[22] the market data following price decontrol suggest something a bit more complex.

The General Accounting Office issued three major surveys of cable rates following federal decontrol. In their 1989 report, it was found that the rates for the most popular basic cable tier rose 19% faster than inflation in the first 23 months of deregulation.[23] While this has been used in the legislative debate and legal literature to suggest that cable enjoys an unregulated monopoly pricing situation due to the 1984 Act (Allard, 1993b), the rest of the story leads elsewhere.

In Table 7.2 it is possible to compare the monthly prices of the most popular basic cable packages for systems that were deregulated by the 1984 Act with those that were already deregulated (24% of the sample fell into this category). Prices rose just 4% faster in the newly deregulated subset.[24] If the systems that were deregulated to begin with are used as a control group, which seems appropriate, this differential appears to be a measure of the marginal price impact of deregulation, before accounting for possible quality changes.[25]

Quality enhancements are theoretically probable (Leffler, 1982) and are strongly suggested by the data. The move to deregulation coincided with a large expansion of capacity, such that the real price-per-channel was virtually unchanged. Of course, if marginal channels are of little value to consumers, then this statistic will not fully compensate for the price rise of the total package. In this instance, subscription rates would be expected to fall. Yet, penetration levels rose from 55.5% to 57.1%[26] in the immediate postderegulation period (Table 7.2) and continued rising in subsequent years.[27] Indeed, the pattern of cable penetration, as displayed in Figs. 7.1 and 7.2, suggests that output growth rose after deregulation and moderated with the advent of reregulation.

TABLE 7.2
1989 GAO Basic Cable Rate Survey

	12/1/86	10/31/88	%Δ
Most Popular Tier Price (All Systems)	$11.70	$14.77	26 (19[a])
Price in Newly Deregulated Systems	$11.58	$14.76	27.4
Price in Systems Already Deregulated in 1986	$12.03	$14.90	23.8
Number of Channels (All Systems)	26.6	32.1	20.7
Price per channel (All Systems)	$.44	$.46	−1.4[a]
HBO Price	$10.46	$10.31	−7.0[a]
Price for a Package of Two Premium Channels	$18.64	$17.82	−9.8[a]
Revenue per subscriber	$21.58	$24.68	7.9[a]
Penetration	.555	.571	2.8
Percent of systems offering just one tier	76.5	84.1	10.0

Note. All dollar figures are monthly.
[a]Inflation adjusted by the GNP implicit price deflator.

It is important to note that cable prices also increased significantly faster than
the Consumer Price Index (CPI) in 1989–1991.[28] Although reregulation advocates used such evidence to insist that cable was exercising monopoly power,
the logic is flawed. Monopoly price adjustments following deregulation would
be expected to result in immediate (or relatively short-term) price increases.
Five years after the fact, however, hefty annual increases were still in evidence.
This suggests that demand continued to shift outward, presumably due to increases in program quality. This is bolstered by the fact that output (measured
by penetration) does not appear to have been restricted.

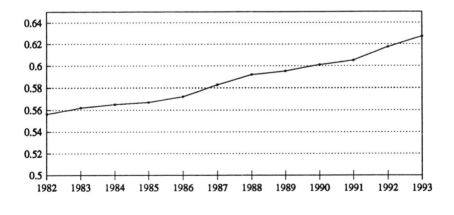

FIG. 7.1. U.S. cable penetration rates (1982–1993). Data are from Paul Kagan
Associates, March 1994.

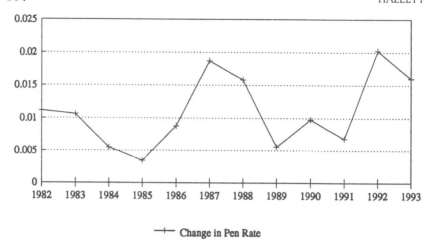

FIG. 7.2. Change in U.S. penetration rates (1982–1993). Data are from Paul Kagan Associates, March 1994.

Indeed, further evidence suggests consumer choice has improved not just in numbers of channels per system but in the quality of the programming on existing channels. Cable system expenditures for basic cable programming amounted (nationally) to $255 million annually ($8.12 per subscriber) prior to deregulation in 1983; by 1992, operators were spending nearly $2 billion (or $35.14 per subscriber) on these key inputs (see Table 7.3). On a per-subscriber basis, this constituted a nominal increase in program costs of 333%.

Quality may be more definitively evaluated, perhaps, by examining viewing shares. This alternative measure of output takes into account inframarginal consumers who continue to subscribe to cable at higher prices. In Table 7.4, market share growth for basic cable networks appears to have accelerated, if anything, in the postderegulation environment.

Data for cable households, which abstract from the increase in the number of households subscribing to cable, reveal that the all-day average viewing share

TABLE 7.3 (Part 1)
Cable Program Expenditures and Internal Subscriber Growth, 1983–1993

	1980	1981	1982	1983	1984	1985	1986
New Homes Passed (mil.)	5.6	6.9	7.7	4.7	4.2	4.2	4.7
Penetration of new HP	30	30	30	30	30	30	30
Internal sub growth (%)	9.0	7.5	8.0	8.0	4.5	3.2	4.1
Basic program spending ($mil.)	234[a]	n.a.	n.a.	255	325	368	496
% change year-on-year	n.a.	n.a.	n.a.	n.a.	27	13	35
Spending per sub ($/yr.)	n.a.	n.a.	n.a.	8.12	9.50	10.04	12.50
% change year-on-year	n.a.	n.a.	n.a.	n.a.	6	7	28

TABLE 7.3 (Part 2)
Cable Program Expenditures and Internal Subscriber Growth, 1987–1993

	1987	1988	1989	1990	1991	1992	1993
New Homes Passed (mil.)	3.7	4.1	5.6	3.2	2.4	2.2	2.0
Penetration of new HP	30	33	36	39	40	40	40
Internal sub growth (%)	4.2	3.7	3.2	2.3	1.5	1.5	1.6
Basic program spending ($mil.)	572	739	1,006	1,410	1,720	1,940	2,189
% change year-on-year	15	29	36	40	22	13	13
Spending per sub ($/yr.)	13.43	16.17	20.41	27.27	32.15	35.14	38.47
% change year-on-year	(7)	20	26	34	18	9	10

Source: Paul Kagan Associates, *Marketing New Media* (15 March, 1993), p. 1, augmented and updated by research request to Kim Weill at Paul Kagan Associates (February 1994).
[a]National Cable Television Association, "Growth of Cable Television," factsheet distributed at June 1993 annual convention in San Francisco.

of basic cable networks was 17% in 1983–1984, the last season prior to the 1984 Cable Act. This share more than doubled, to 35%, by 1990–1991 (see Table 7.4). Whether this constituted an improvement from trend is difficult to discern. Paul Kagan Associates provides audience ratings for cable networks only back to 1983. More importantly, due to the deregulation of cable program content by the courts and the FCC in the mid- to late 1970s,[29] there was a bubble in network growth and viewership about this time. The pattern revealed in Table 7.1 is informative. Two incidents of deregulation appear to boost cable network formation: The first (eliminating controls on programming) led to a boom in 1978–1981 (when national satellite video nets increase from 9 to 38); the second—federal rate decontrol—promoted a boom in 1984–1987 (when basic networks, specifically, jumped from 26 to 52). It appears safe to conclude that cable television was becoming both more diverse and more popular with viewers following each round of deregulation.

TABLE 7.4
Broadcast and Cable Viewing Shares in U.S. Cable TV Households

Category	'83/'84	'84/'85	'85/'86	'86/'87	'90/'91	'91/'92	'92/'93
B'cast Net. Affiliates	58	56	56	53	46	47	46
Independent B'cast Stns.	17	17	17	17	17	16	17
Public Stations	3	3	3	3	2	3	3
Basic Cable Networks	17	19	19	23	35	35	36
Pay Services	11	11	10	10	9	8	8

Source: Nielsen Ratings reported in National Cable Television Association, *Cable Television Developments* (April 1994), p. 5A. Note that shares may sum to more than 100 due to the use of multiple TV sets within homes.

An entirely unsurprising result of deregulation is also observed with the elimination of tiering. Under the price regulations imposed by local governments, firms were able to shift desirable channels off the regulated package and into "expanded basic"—at generally unregulated prices. When decontrol came, industry analysts instantly proclaimed a shift away from tiering.[30] Expanded basic subscribership reached a peak in 1986 of 6.6 million subscribers, declined to just 3.8 million subscribers in 1989, and then rose rapidly again with the threat of reregulation to 12.2 million in 1990 and 15.0 million in 1992.[31] Similarly, it is found in the GAO data that the number of firms offering tiers beyond the standard package quickly fell from 23.5% of the systems surveyed to 15.9% (see Table 7.2) before heading up to 41.4% in 1991 amid fears of reregulation (GAO, 1991, p. 2). One can additionally see a reduction in pay prices (always deregulated). The picture that emerges is that cable operators responded to free-market pricing by consolidating packages, adding channels, upgrading program quality (spending more per channel), and dropping premium rates.[32]

Cable-asset values rose dramatically in the postderegulation period. Table 7.5 demonstrates virtually a doubling of the average capital value of a cable system over the 1985 to 1988 period. It is likely that deregulation had a positive impact on sales prices, but the deliberate, year-to-year, asset-value increases appear partially due to external factors such as the declining real-interest rates that were driving the overall stock market boom.[33] Zupan (1989, p. 409) noted that the cable industry stock index rose by 16% in the 2 months following the 1984 Cable Act, compared with an 8% rise in the S&P 500. Why it would take as long as 2 months to capitalize the benefits from proindustry legislation remains problematic. But, if one does ascribe the entire above-market gain of 8% to

TABLE 7.5
Sales Price Data for U.S. Cable Systems, 1982–1992

Year	# Systems	Total Subs	$Value/Home Passed	$Value/Subscriber
1982	212	934,071	486	922
1983	256	2,631,190	554	1,026
1984	295	3,023,144	520	948
1985	356	7,992,899	546	1,008
1986	620	6,797,164	733	1,339
1987	498	6,506,466	946	1,723
1988	596	7,596,344	1,162	2,003
1989	379	5,951,353	1,255	2,291
1990	105	531,207	1,277	2,049
1991	111	4,523,433	961	1,795
1992	97	1,876,754	1,023	1,768
1993	97	5,472,668	1,264	2,124

Source: Paul Kagan Associates, Cable TV Investor (25 January, 1993), p. 13; (12 February, 1993), p. 6; and (28 February, 1994), p. 12.

deregulation, this is still only a small fraction of the capitalization gains occurring postderegulation.

The slow, gradual rise in cable asset values during the mid- to late 1980s suggests that, even if deregulation attributed to the value changes, something more subtle than a relaxation of price controls was the cause. Consumer acceptance of a higher-quality, higher-priced cable package—as witnessed with the introduction of new services and higher rates in the 1987–1989 period—is the more compelling explanation. In eliminating the costs (including cross-subsidies) and riskiness associated with regulatory approval of rate increases, deregulation may have been indirectly responsible for this take-off in asset values. Yet this explanation has distinct consumer welfare implications from the view that price decontrol led to enhanced exercise of market power.

5. FEDERAL REREGULATION (1992)

The Cable Act of 1984 concluded a decade of deregulation. The result was a golden age for cable. Nationwide, the proportion of subscribing households tripled between 1976 and 1988; by 1993, 96% of U.S. homes were passed by cable with 60% of all households subscribing. The cable industry was collecting over $20 billion in annual revenues, system values doubled, and cable-only networks were accounting for nearly one third of all U.S. TV viewing.[34] Ironically, this marketplace success exposed the cable industry to political risks it had previously surmounted. A coalition led by broadcasting interests placed cable reregulation onto the policy table. Congress began debating reregulation of cable by 1988 and got seriously close to enacting legislation in October 1990.[35] By October 1992, a bill that included reregulation of cable television rates was enacted over President Bush's veto—the only instance of an override during the Bush Administration.

The Cable Consumer Protection and Competition Act of 1992 was a comprehensive measure that included the following[36]:

* Reversed 1984 Cable Act by reregulating basic cable rates (limited basic) and cable programming services—expanded basic tiers (i.e., everything up to á la carte or pay-per-view offerings). Directed the FCC to define the terms of regulation. In April 1993, the Commission decided to require cable systems to roll back basic prices up to 10%, on a per-channel basis. In February 1994 it upped the rollback to 17%.

* Allowed local TV broadcasters to charge cable systems for retransmission of their signals. Since the 1976 Copyright Act, broadcasters had been uncompensated. When the property-rights switch officially arrived in October 1993, however, cable companies refused to pay a fee to broadcasters, and the great majority of retransmission deals were consummated with no more than limited

in-kind (such as marketing cross-promotions) consideration. (Additionally, some new cable networks were created by broadcasters that will receive license fees in deals struck in lieu of retransmission payments.)

* Alternatively, gave broadcasters option of "must carry"—electing to give up any retransmission fees in exchange for (mandatory) carriage on local cable systems.[37] Most smaller, nonnetwork television broadcasters selected this option.

* Modest procompetition rules (e.g., program access for competitive video providers, uniform pricing by local cable systems that had been discriminating by dramatically underpricing new entrants, and directing cities to issue competitive franchises). Some rules were undermined by other provisions (e.g., the elimination of monetary damages against municipalities that unreasonably refuse to grant competitive cable licenses).

* No change in the 1984 telco-cable cross-ownership ban.[38]

* At least 24 FCC rulemakings and reports mandated. Issues include interior wiring rules, horizontal concentration, vertical integration, controls on indecent or obscene programming, tier buy-through provisions, customer service, small cable system regulatory exemptions, cost-of-service adjustments to price regulations, and so on.

The evidence concerning the 1992 Cable Act's impact on product quality is difficult to quantify at this early stage. But the outlines of the industry's reaction to rate controls are already clear and are consistent with theoretical predictions. Cable suppliers are lowering the quality of the regulated tiers of service and shifting program services to unregulated status. Overall, there are no obvious indicators that regulation is promoting an increase in nationwide cable-viewing shares or in the penetration rate of cable subscribership, as would be expected if rates were reduced and quality stayed constant or increased (or fell by a lesser proportion than the decline in prices, as judged by the marginal consumer). There are, however, two interesting footnotes that describe how firms taking evasive maneuvers to escape price controls may improve product quality, although in ways that may entail significant social costs.

The clearest evidence is that there is no pronounced increase in basic penetration (see Figs. 7.1 and 7.2). Further, viewing shares for basic cable channels appear to have leveled off over the year or two preceding cable regulation.[39] A similar pattern appears in the internal growth rate of cable subscribers, a measure that largely abstracts growth gains due to new construction (see Table 7.3). This rate of growth fell from the 1984–1990 period, when it averaged 3.4% annually, to just 1.5% in the 1990–1993 period.[40] Basic cable program expenditures per subscriber did increase by 9% in 1992 and by 10% in 1993 (see Table 7.3), but these increases are significantly below the 26% and 34% increases posted in 1989 and 1990. Whether this apparent break in trend was related to gathering momentum for reregulation cannot be proven by the data, but it is consistent with the explanation of the impact of reregulation given in the trade press and as reflected in the formal public policy debate, as shown later.[41]

At this point, a clearer picture may be had by examining the retiering strategies being employed to fundamentally alter the delivery of basic cable service by restructuring premium (unregulated) offerings and the political positions actually taken by affected interest groups vis-à-vis cable rate regulation. There is an abundance of anecdotal evidence concerning (a) adjustments that cable systems are undertaking in response to rate reregulation, amply covered by the trade press; and (b) the official positions taken by the various interest groups during the public debate concerning reregulation. But first it is necessary to describe the extent and form of the rate-regulation scheme itself.

Rates Under Reregulation[42]

The process by which cable rates are regulated under the 1992 Act is complex. The administrative structure created by Congress relied heavily on the Federal Communications Commission to determine exactly how controls would be instituted and what the regulations would consist of. Only the bare outlines were given by the Act itself.

The Act mandated that there would be two levels of rate regulation: the first for basic service, a tier including off-air broadcast signals as well as public access and government channels (popularly referred to as "limited basic"); the second for all higher tiers of "programming services."[43] The first level of service would be primarily regulated by local governments, although the FCC was instructed to create regulatory guidelines and to officially certify local franchising authorities before they were allowed to carry out rate regulation.[44] The price standard created by the FCC under the Act is that consumers are not supposed to be charged more than they would be if, in fact, "effective competition" prevailed in their community. The second level of regulation would be conducted directly by the FCC in response to complaints filed by local citizens or by the certified regulatory boards. The FCC is required to regulate if, upon such formal complaint, local cable rates are found *unreasonable*. The Act lists at least six factors to be taken into account by the FCC in defining this term.

On April 1, 1993, the Federal Communications Commission announced that it would freeze all tiers of basic cable at their levels as of September 30, 1992, and institute rate rollbacks as of September 1, 1993, by as much as 10%.[45] Such rate regulations were to be instituted on a per-channel basis. Each system would determine, according to its size and number of channels, "benchmark rates" for its basic programming services according to tables published by the Commission. It would then be allowed to charge either the benchmark rate or 90% of its per-channel rate as of September 30, 1992 (adjusted for inflation up to the present). This format applied to both basic and programming services.[46]

The 10% (maximum) rate rollback was based on the FCC's estimation of the difference between monopoly and competitive cable prices, but it employed a troublesome definition of the latter. The 1992 Act left it to the FCC to define its own "reasonable" rate standard, but did—in another context[47]—define *effective competition* in three ways:

Type A: systems serving under 30% of the homes in their franchise areas.

Type B: systems serving markets in which a second multichannel video operator can serve at least 50% of the households and does serve (as subscribers) at least 15%.

Type C: systems that are municipally owned or are private systems competing with a municipally owned system that passes 50% of the homes in the franchise area.

The FCC elected to borrow this statutory definition of effective competition and to estimate a price equation of the following form:

$$LNP_i = \alpha + \beta_1(ABC_i) + \beta_2(RECIPSUB_i) + \beta_3(LNCHAN_i) + \beta_4(LNSAT_i) + e_i$$

where LNP = natural logarithm of the composite price per channel for up to three tiers of service, weighted and adjusted to exclude franchise fees and include equipment and other subscriber charges as described previously; ABC = 1 if the community unit belongs to one of the categories comprising the statutory definition of effective competition, as described earlier, and ABC = 0 otherwise; RECIPSUB = 1 per number of households subscribing to the cable system; LNCHAN = natural logarithm of the number of channels in use in the tiers of service examined; and LNSAT = natural logarithm of the number of satellite-delivered channels in the tiers of service examined (FCC, 1993, Appendix E, p. 12).

This price equation was estimated using data on 377 cable systems, of which 79 were Type A, 46 Type B, and 16 Type C. As a group (ABC), the dummy coefficient equals −0.0939 and is significant at 99%. The Commission rounded this up to 10% and identified it as the percentage price discount associated with competitive systems. This was controversial, and a further round of public comment was undertaken by the Commission,[48] although the Commission declined to change its methods prior to the September 1993 rollback. When run as separate dummy samples in the FCC's model, the three distinct definitions of competition produce coefficients of +9.2% (A), −22.1% (B), and −38.7% (C), all of which are statistically significant at 99%. In other words, the A group demonstrated prices above those typically found in monopoly systems. The inclusion of these systems dilutes the price-lowering impact of actual head-to-head compilation such that only about a 10% differential is observed.[49] This estimation procedure formed the basis of the Commission's rate rollback in September 1993, although the FCC is reported to be considering a reexamination of the 10% figure.[50]

.

Retiering, Repricing, and Restructuring Cable Program Services

The low-cost avoidance method with which to deal with cable rate controls remains retiering. The twist regarding the 1992 Cable Act is that policymakers were aware of this escape route and included legislative provisions that were designed to deal with it, giving the FCC authority to regulate all tiers of "cable programming services"[51] pursuant to complaints from either local government officials or cable customers. This has increased the distance that cable companies must traverse to spring basic cable networks free of rate regulation. It has led to two forms of retiering: (a) adjustments between limited basic and expanded basic tiers, which largely concerned operators preparing for the April 1, 1993 "freeze," and (b) a shift of high-quality cable networks off basic altogether (either dropping channels or moving them to premium status).

Limited basic service is being raised in price in most markets, whereas higher tiers are being lowered—such that the net result for a customer subscriber to both is naught. (This is ironic, in that such basic tiers were traditionally created as low-cost "lifeline" services.) Cable television subscribers in Hollywood, CA, for example, found the restructuring worked as shown in Table 7.6.

This shift was prompted by concerns that regulations going into effect in April 1993 would freeze basic cable rates at artificially low levels. More complicated restructuring also took place. The general pattern: prices for limited basic rise; the incremental charge for expanded basic falls; charges for additional outlets and remote controls fall; other incidental charges increase.

For instance, cable companies are attaching "cost-based" installation charges. What had been a free hook-up for Albertville, AL subscribers is now $10.05. Also, companies believe they actually can charge for additional outlets if pay channels (such as HBO or Playboy) are on the additional television. A Houston system is still able to charge $3.95 per month for an additional outlet this way.

TABLE 7.6
Cable Rate Changes in Hollywood/Wilshire Franchise

	Current Price	New Price
Basic Broadcast Service	$2.10	$9.85
Standard Service	$19.90	$13.50
Remote	$2.10	$.75
Basic Broadcast Service plus Standard Service and Remote Control Package	$24.10	$24.10
Additional Outlet with Remote Control Package	$7.88	$7.88
—Additional Outlet	$5.78	$7.13
—Additional Outlet Remote Control	$2.10	$.75

Source: Letter to Continental Cablevision subscribers in Hollywood/Wilshire Los Angeles franchise announcing April 1, 1993 rate schedule.

And many systems are tacking new line items onto customers' bills; Toledo, OH cable customers used to receive converters and home service calls at no additional charge. Now converter boxes are billed at cost—$2.54 per month—as is home wiring—31 cents a month.

Systems are experimenting with various combinations to see how to make the new regulated package revenue-neutral by shifting basic programming services to unregulated status. The strategy can either entail raising premium channel prices (unregulated) and marketing them more intensely, or putting newer, cheaper, less watched channels on basic tiers and shifting popular networks to à la carte status. A system in San Antonio, for instance, is splitting TBS and WGN off into per-channel status: $1 and 50 cents, respectively, or $1.25 a month for both. E! (part time), VH-1, and the Comedy Channel were added back into basic. The net result was that rates dropped 17 cents per month.

Many systems actually increased prices for basic cable service: "In Paragon Cable's Manhattan system, the overall basic-plus-standard rates will rise from $22.95 to $23.65, which includes a $3.32 converter fee and a 20-cent charge for a remote."[52] (Rates also rose from $22.95 to $23.58 in Time Warner's New York City system.) In the Staten Island Paragon system, six channels (Madison Square Garden Network, MSG2, TBS, Discovery, AMC, and the Cartoon Network) were severed from the basic package and put on à la carte basis. Costing from 50 cents to $2.00 per channel, the package sells for $3. Even with the à la carte tier, however, the basic package dropped $1.60 per month.[53]

The degree to which nominal prices have gone up or down following the September 1993 implementation of rate regulation is not well understood, as shown earlier. The picture is even more complicated, obviously, when quality changes are accounted for. Tele-Communications, Inc., the largest cable operator (serving nearly 20% of U.S. subscribers), was chagrined when a memo written by one of its vice presidents and sent to over 500 system managers was leaked to the *Washington Post* in November 1993. The memo outlined how the company could raise prices for "downgrades, upgrades, service calls and VCR hook-ups," as they were unregulated under the new rules. "We cannot be dissuaded from the charges simply because customers object," wrote the TCI executive. "It will take a while, but they'll get used to it." His conclusion was explosive: "The best news of all is we can blame it on reregulation and the government now. Let's take advantage of it!"[54]

Another affect of reregulation appears to entail the substitution of cheaper and/or lower quality programming for existing cable networks. C-Span, a high-quality (if inexpensive) public affairs network, suffered losses mounting to 1,000,000 subscribers (either dropped entirely or reduced to part-time carriage) on "Sept. 1 [1993] as systems retiered rates and channel line-ups."[55] Broadcast stations and home shopping outlets are convenient stations to add to basic packages both because they reduce the cost per channel and they comply with the "must carry" rules contained in the 1992 Act. Home shopping cable networks actually pay for carriage, giving cable operators added economic incentive to add such channels. They also tend to dilute quality, however, or the cable operator would presumably have been offering such programming preregulation.

The same incentive, however, may have beneficial impacts on basic cable quality by prompting operators to add channels, thereby giving some upstart networks additional audience coverage.[56] The way the FCC rate regulations have been crafted also has led most systems to lower charges for additional outlets, which may in turn increase audience share of basic cable networks.[57] With enhanced advertising revenue streams, the quality of these channels could rise over time. Ultimately, however, channels that are added simply to alleviate binding price constraints could themselves be replaced by programming that is cheaper still.[58] Moreover, the overall impact on the quality of cable network programming does not appear to be positive. Not only have some cable networks lost significant carriage, cable audience shares are not increasing relative to trend, and both producers and programmers have tended to strongly oppose rate regulation, as discussed later. It is difficult to conclude that such ad-hoc mechanisms to water down rates represent a long-lived equilibrium.

The ultimate irony may be that reregulation will speed technological change. Over the medium to long term, systems may have greater incentives to change the (regulated) marketing margins altogether by upgrading cable plant to the 500-channel environment. Combined with addressable electronic controls, this will circumvent the regulatory regime almost entirely by shifting to virtual video-on-demand delivery. Regulation will be rendered moot either by adding vast numbers of "Fishbowl Channels"[59] or by taking the entire cable package á la carte.[60]

These incentives to improve product quality may not enhance consumer welfare, however, as they derive from rent-seeking behavior. Even socially useful investments will entail welfare losses if, due to strategic behavior, they are undertaken too soon or with the wrong production function.[61] This is especially important in that enhanced competitiveness appears to be a policy substitute for rate regulation. It is now apparent that market forces are themselves pushing both convergence of technologies across several telecommunications markets and competition between delivery systems. The impact of allowing telephone company competition (which the 1984 Act specifically prohibited and on which the 1992 Act was silent) is but one of a number of procompetitive strategies that could produce a market-driven result producing greater channel capacity and a broader selection of video choices.

Political Coalitions and the 1992 Cable Act

It is perhaps easiest to gauge the impact of cable reregulation from the self-interested positions taken on the issue of cable rate regulation. Employing the assumption that economic interests tend to loyally assert the public policy position consistent with profit maximization, one can examine the key participants in the debate on rate regulation to gain an understanding of its likely effects.

It is straightforward that cable operators vigorously opposed the Cable Act. This does not necessarily imply, as some have asserted, that the Cable Act would have the likely effect of lowering quality-adjusted prices for consumers (Carroll & Lamdin, 1993). Cable interests would reliably oppose added constraints that do not provide offsetting benefits.[62] But constraining profits does not necessarily transfer surplus to consumers. If cable systems lower quality by a sufficient degree, rate regulation can clearly lower consumer surplus (and, of course, industry profits, which cannot increase with price controls in that adding a constraint cannot improve firm pricing decisions). The evidence is clear that the cable industry did oppose rate regulation, going so far as to conduct a national advertising campaign claiming that reregulation would raise consumers' rates.

Far more interesting is the position taken on rate regulation by the owners of cable programming. As a group, cable programmers were strongly opposed to reregulation. They openly stated their fear that via retiering and other operator adjustments reregulation would negatively impact demand for basic networks. New nets, such as the Sci-Fi Channel, were particularly fearful of regulation. They were particularly vulnerable to suppliers' reactions to price controls, either from being pushed off basic into à la carte status or by failing to gain carriage at all.[63] The actual producers of the programs themselves, represented by the Motion Picture Association of America (MPAA), were also strong opponents of reregulation.[64]

If rate controls did, in fact, lower prices charged by the retail distributors of programming, this would increase penetration and, all else equal, raise the demand for software inputs (networks and programs). Moreover, it would increase audience sizes for basic cable networks and increase their ad revenue streams. Their opposition to reregulation indicates that they believed that the quality-adjusted price of cable would increase and the demand for their programming would thereby fall.[65] Their fears regarding reregulation were very quickly realized: "The Cable Act of 1992 has already adversely impacted cable operators. It is causing a virtual freeze in new programming decisions. Cable operators are proceeding very cautiously when it comes to adding new services like the Cartoon Channel and the Sci-Fi Channel because it may prove difficult, if not impossible, to recoup the investment."[66]

Most interesting of all was the position taken by the broadcasters. Long in a competitive position vis-à-vis cable, particularly in policymaking in Washington, DC, the broadcast industry was keenly interested in the Cable Act. In fact, they were the chief interest-group backers of the legislation, funding a nationwide ad campaign promoting the measure. The industry had long pushed cable rate regulation, including arguing forcefully for it in a 1990 FCC proceeding in which no other issues (such as must-carry and retransmission consent) were involved.[67]

The broadcast industry, as a competitor with cable for viewing audiences, would be expected to benefit from measures that raise the quality-adjusted price of cable services, as this would prompt consumers to substitute away from cable programming into broadcast television fare. In promoting rate regulation, the broadcast industry signals its view that quality (as evaluated by consumers) will adjust downward by more than price, causing a migration of viewers in its direction. This could only come as welcome news to broadcasting, an industry that has seen its market share drift inexorably to cable in recent years. When the FCC released a study in 1991 that described this trend in painful detail, a broadcasting trade journal wishfully editorialized: "Congress . . . may well be inclined to follow the report's lead by putting the brakes on cable's expansion—by reregulation of the wired world while the FCC frees up the broadcast universe."[68]

The evidence gleaned from the rent-seeking competition to obtain favorable legislation speaks loudly: Reregulation was expected to reduce quality by at least as much as it lowered price. This would decrease consumer demand for cable-only programming and increase demand for the substitute television product—broadcasting. Nothing that we observe in the early days of reregulation contradicts such expert testimony.

QUALITY AND PRICE REGULATION

The lessons from cable rate regulation have been diagrammed in recent policy regime switches, as we have gone from regulation to deregulation, and back again. The evidence clearly indicates that operators will adjust service quality as predicted by microeconomics. They will attempt to circumvent controls by reducing quality. In cable markets this is done with particular ease. Programming inputs are highly mobile, and suppliers' demands for software are highly elastic with respect to the prevailing regulatory regime. Satellite networks flourished after deregulation in the mid-1980s and are very nervous about their fate after reregulation in the 1990s. Most fundamentally, cable operators enjoy constitutional protections in their choice of viewing fare supplied, and even if regulators could successfully monitor the price of a given set of channels, they are barred by law from controlling the value of the programming provided thereon.

The evidence from deregulation in both California in the early 1980s and nationwide over the late 1980s indicates that, while prices rose, quality adjusted upward so as to entirely offset such changes. The performance of cable penetration and basic cable viewer ratings indicate that output expanded rather than contracted under decontrol. The current experience with reregulation appears to substantiate this analysis, particularly as how broadcasters—selling the substitute product—are most anxious for cable companies to succumb to the hand of reregulation. This is either curiously altruistic or an affirmation of the view that such price controls raise the effective price of quality-adjusted cable service.

It would be ironic, however, if the best evasive maneuver employed by regulated cable companies turns out to be a hastened leap into the next generation of technology. In that event, one could argue that the incentives for cable firms to avoid rate regulation were so strong that they abandoned their traditional market entirely, creating a substantially new product space as a safe haven. New and improved video service on the information superhighway might well exhibit higher quality, yet the firms supplying it might well be able to exploit even greater degrees of market power unless new forms of competition are brought to bear. Then again, such competition could—by all outward signs—have delivered the next generation of technology to the subscriber's door even faster had the prohibitions against it been relaxed to begin with. Using regulation to encourage quality-enhancing evasion seems a rather circuitous and danger-filled path to a long-run optimum.

ACKNOWLEDGMENTS

The author thanks Lorraine Egan and Martin Morse Wooster for reliable research assistance.

ENDNOTES

1. See Cheung, 1974, for a general theory of price regulation, and Hazlett, 1991, for an analysis of how price controls affect cable television quality.
2. This mobility assumption applies to factor inputs and the divisibility of outputs.
3. In his 1974 article, "A Theory of Price Control," Steven Cheung used the rent control market to establish what he believes to be a general paradigm for price (or other) regulation. The theory revolves around how buyers and sellers will attempt to claim or dissipate the economic rents that become, in essence, common property when binding controls create excess demand. That Cheung focused on apartment rent controls led him to skip the straightforward point that a supplier (landlord) can reclaim lost property rights by withdrawing input expense. That withdrawable operating costs are such a small portion of the apartment supply function apparently kept this insight hidden. It can be included in his property rights framework, however, by noting that the withdrawal of inputs that are worth at least their marginal cost to consumers is a wasteful rent-seeking dissipation: To recoup some of their rents lost from price controls, suppliers are willing to curtail socially efficient investments.
4. The difficulty in measuring cable program quality is such that economists who have undertaken this task have used such measures as "total channels offered to subscribers" as a proxy (see Otsuka, 1993). That such an approach is problematic is obvious to anyone who has flicked a cable television remote control. Not only are not all channels created equally, both cable firms and cable regulators have historically had incentives to cross-subsidize particular channels and even channel capacity itself (see Hazlett, 1986).
5. Since the late 1970s, cable operators have won a series of landmark cases establishing their First Amendment rights as "electronic publishers." These bar local officials from exercising authority over what channels are carried or the shows such channels carry. Recently, this status as Constitutionally protected publishers received a large boost when it was extended to a telephone company attempting to compete in cable. The federal court decision found that Bell Atlantic, a regional telephone company, had a First Amendment right to provide transport of

video signals and to own the programming that was provided directly to subscribers. (See *The Chesapeake and Potomac Telephone Company of Virginia et al. v. United States of America*, U.S. District Court for the Eastern District of Virginia, No. 92-1751-A [24 August, 1993]).

6. The problem of attempting to control two outcomes (price and quality) with but one policy instrument may not be simply remedied by reregulation. The Constitutional constraint presumably remains. A recent Federal Trade Commission report identified the statutory problem by noting that: "The 1984 Cable Act . . . may make it more difficult for local governments to threaten non-renewal. Section 626(c)(1) limits the criteria that the government may use in deciding to not renew an operator's franchise. This decision may not be based on the prices charged by the operator, nor on 'the mix, quality, or level of cable services or other services provided over the system.' The fact that cities cannot use them in renewal decisions likely vitiates the usefulness of the franchise bidding process as a regulatory mechanism" (FTC, 1990, p. 34). Although the report cited First Amendment case law in cable, the FTC appeared to be confused about the ability of Congress to reregulate should it choose to (as it did in 1992). Yet, no such confusion is warranted. First Amendment protections for cable operators are strong and likely to increase in future years.

7. Currently, nine states assume some authority for cable TV regulation: Alaska, Connecticut, Delaware, Hawaii, Massachusetts, New Jersey, New York, Rhode Island, and Vermont. ("Governors Urge Restraint in Preempting States in Development of National Telecom Principles," *The Cable-Telco Report* [1 August, 1994], p. 17.)

8. The FCC is planning to allow systems to charge higher-than-capped prices via a cost-of-service showing as a safety valve measure. Such procedures have not yet been crafted but are expressly created as special-case exemptions from the general rules.

9. GAO, 1991.

10. Broadcast stations that are distributed nationwide to other cable systems (such as WOR, WGN, or WTBS) are called *superstations* and are generally counted as cable networks.

11. They continue to be exempted from price controls in the 1992 Cable Act.

12. According to FCC Commissioner James Quello (Allard, 1993b, p. 107).

13. When examining shifts in demand for cable, as measured by such indices as subscriber penetration rates, however, consumer preferences will reflect these dimensions of quality.

14. The analysis for monopsony buyers, obviously, mirrors the analysis for monopoly sellers.

15. This section relies heavily on Hazlett, 1991.

16. Penetration here means subscribers per homes passed. This is a measure of output adjusted for system size. Within the context of the regulations, it was thought that systems with greater than 70% penetration were located in areas in which cable was more of a necessity.

17. Differences in the price changes of newly deregulated firms over a 2-year period reveal a fly-up in rates of just 5.58%. Penetration results are similar as for the 1-year experience.

18. Direct measurements of the quality of cable service (e.g., cable channel ratings for California cable systems) are not available.

19. This does not mean that the rate controls were meaningless or foolishly imposed by the political system. Their importance was in helping local officials and interest groups enforce the rent distribution schemes that had been part of the original franchise agreements. All else equal, firms would rather price in an unconstrained environment, charging a high monopoly price for a high-quality cable package. They will, if constrained by price controls, charge a low monopoly price for a low-quality cable package. But that is a second-best alternative, as indicated by revealed preference on the supply side.

20. Typical cable franchises are awarded for durations of 15–20 years. For a discussion of the franchising process, see Hazlett, 1986.

21. The then-president of the National Cable Television Association, James Mooney, obviously agreed with this assessment. "Hanging in Mooney's office is a copy of the Cable Communications Policy Act of 1984, which deregulated cable, enriched cable operators and contributed to the cable programming boon of the 1980's. It is a tribute to Mooney's legislative prowess. The act

is the only major amendment to the Communications Act of 1934—the basic charter of communications law. Mooney would like to keep it that way" (Harry A. Jessell, "Mooney: Rereg No Sure Thing," *Broadcasting* [4 May, 1992], p. 15).

22. "Why have basic cable television rates shot up about three times inflation in the past two years, after more than a decade of stable prices 30% to 40% below inflation? The answer is simple: Despite the fact that virtually no consumers have more than one cable company to choose service from, Congress allowed the Federal Communications Commission to deregulate cable pricing a few years ago. With growing evidence that the cable industry is price-gouging video consumers, Congress must correct its mistake and put a lid on cable rates" (Gene Kimmelman, "Slam a Lid on TV Cable Rates," *Cleveland Plain Dealer* [6 June, 1990]; [Kimmelman is executive director of the Consumer Federation of America]).

23. Although the Cable Act passed in 1984, it set the following timetable for rate decontrol: 1985: 5% rate increased allowed; 1986: 5% rate increase allowed; 29 December, 1986: price decontrol in any cable system deemed "effectively competitive" by the Federal Communications Commission. The FCC so defined a cable system serving a community in which just three over-the-air broadcast signals could be received. This deregulated 97% of U.S. cable systems.

24. The GAO does not publish standard deviations to accompany its mean values, ruling out tests of statistical significance.

25. A 1990 GAO survey found that in the November 1986–December 1989 period this differential grew to 15% (47% vs. 32%). Although these figures are likely to be distorted by measuring the lowest priced basic tier (which includes "sham" rates with virtually no subscribership) instead of most popular tier prices, the 4% difference found in the shorter period and the 15% difference found in the longer period bound the 10% difference found in the California data.

26. Overall cable subscribership increased 15% during the 19-month period, but much of this growth was due to new cable plant being constructed. A rise in penetration rates tends to adjust for plant size.

27. There is some disagreement over cable penetration (and other) numbers, but the alternative sources appear to agree on their trend. According to an FCC study, average basic cable penetration was 55% in 1980, 56.7% in 1985, and 61.4% in 1990 (Setzer & Levy, 1991, p. 68). The GAO penetration results are also roughly consistent with Paul Kagan's numbers; in 1986, mean MSO (multiple system operator) penetration was 57.4%, rising in 1988 to 58.5%. Paul Kagan Associates, *The Cable TV Financial Databook* ([June 1987; p. 55], [June 1989; p. 64]).

28. For instance, basic cable prices increased 15% between December 1989 and April 1991, according to a GAO survey. From the December, 1986 deregulation (4½ years previous), the price of the most popular basic programming tier was found to have increased 61% in nominal terms or 36.5% when adjusted for inflation. Again, the real price change per-channel was virtually nil. As the number of channels received on this tier rose from 27 to 35, the real price increase per channel was 5.3%, or 1.2% annually (see GAO, 1991, pp. 2, 5).

29. In the 1960s, the Federal Communications Commission had enacted anti-cable rules in order to protect television broadcasters from competition (see Besen & Crandall, 1981).

30. Paul Kagan wrote: "Expanded basic, which was originally intended to circumvent basic rate restrictions, will be a casualty of deregulation" (Paul Kagan Associates, *The Cable TV Financial Databook* [June 1987], p. 10).

31. Paul Kagan Associates, *Cable TV Investor* (12 February, 1993), p. 5. The 1992 Cable Act was actually the culmination of years of debate in Congress over reregulation; a debate that featured a flurry of new bills in 1990. As the GAO wrote: "Some of the legislative proposals introduced in 1990 would have generally restricted rate regulation to only the lowest tier" (GAO, 1991, p. 2).

32. This was just what the industry observed at the time: "Operators took the opportunity to repackage and remarket services by emphasizing basic's value and cutting pay prices. Despite double-digit basic price hikes in early 1987, the industry found little if any price resistance from subscribers. New services and original programming are easing the transition to higher rates

and have attracted new subscribers" (Paul Kagan Associates, *The Cable TV Financial Databook* [June, 1987], p. 10).

33. This interest-rate sensitivity appears obvious in hindsight. With the credit crunch and HLT (highly leveraged transactions) restrictions placed on cable financing by federal regulators in late 1989, system prices dropped sharply.

34. In the 1991–1992 television season, basic cable averaged a 24% viewing share, whereas pay channels averaged (a combined) 6% (NCTA, June, 1993, p. 5A).

35. In May 1990, a leading cable analyst wrote that cable stocks plunged nearly 20% in the last quarter of 1989 due to three factors, one of which was "proposed cable rate reregulation advanced by the U.S. Senate" (Paul Kagan Associates, 1990, p. 6). (The other two were the collapse of the junk bond market, and government banking restrictions on highly levered transactions.) The reregulation threat came and went—and came for good—over the next 3 years.

36. See also Allard, 1993a, 1993b; Hazlett, 1993b, 1993c.

37. "Must-carry" rules had been struck down in 1985 and again in 1987 by U.S. courts as violations of the cable system operators' First Amendment rights to select their own programming. In 1993, the U.S. District Court of Appeals (D.C. Circuit) surprised many industry analysts by approving the "must-carry" rules contained in the 1992 Act (*Turner Broadcasting System, Inc. v. FCC*, No. 92-2247 [D.C. Cir.; 8 April, 1993]). The U.S. Supreme Court over-ruled this decision, remanding the case to a district court for further fact-finding (*Turner Broadcasting System, Inc. v. FCC* [S. Ct., No. 93-44; 27 June, 1994]).

38. The ban on telco entry into video has since been found unconstitutional by a federal court in the suit filed by Bell Atlantic (referenced in endnotes). The issue is being litigated both on appeal and in actions filed in other jurisdictions by each of the remaining six Regional Bell Operating Companies, as well as GTE and Southern New England Telephone. U.S. West has also received a favorable opinion from the U.S. District Court in Seattle (*U.S. West, et al. v FCC*, No. C93-1523R, "Order Granting Plaintiffs' Motion for Summary Judgment and Denying Defendants' Motion for Summary Judgment," [U.S. District Court, Western District of Washington; 15 June, 1994]).

39. Looking at cable households only, average all-day viewing shares continued to increase in 1992 and 1993, but at a considerably lessened pace from that seen in 1986–1990. Over the 3 years immediately following deregulation, viewing shares increased from 23 to 35, or 16.0% annually. The annual increase amounted to just 1.4% between 1990 and 1992 (see Table 7.4).

40. This is calculated geometrically from the annual growth rates shown in Table 7.3. Note that the internal growth rate measures the percentage increase in new subscribers on existing plant. It expands both due to new housing (which fills in on existing cable) and due to penetration increases. Hence, the housing slowdown associated with the 1990–1991 recession undoubtedly slowed internal growth independent of any shift in cable demand. Importantly, however, the 1992 reregulation does not appear to shift the rate upward, as would be implied if the rate freeze and subsequent "rollback" had significantly lowered quality-adjusted prices relative to trend.

41. A marketing survey appears to show that consumers' "perceived values" for cable programming were declining over the 1991–1993 period. A firm that surveys 1,000 cable customers annually, asking them to explicitly put a value on the top 20 basic cable networks, found average values declined 8% between 1992 and 1993 (and 38% between 1991 and 1993). What to conclude from this is uncertain, however, as the reliability of this evaluation method is suspect in that it does not measure actual consumer choices (Paul Kagan Associates, *Marketing New Media* [20 December, 1993], p. 2).

42. This section follows Allard, 1993b.

43. Pay-per channel and premium channel services, in which consumers pay for the individual channel or program, were explicitly exempted from price regulation.

44. If the local franchising authority fails certification, the FCC is required to regulate cable rates in the jurisdiction itself.

45. A host of other issues were addressed, and the explanation for these rules explained. The Report and Order was 521 typewritten pages, single spaced. The Notice of Proposed Rulemaking in December 1992 had inspired comment from 176 parties and reply comments from 121.

46. Equipment charges, such as monthly fees for remote controls, additional outlets, converter boxes, and so on, were also controlled, but the rate benchmarks were binding on the overall package including equipment rental. Hence, if equipment charges are reduced, this allows operators to raise monthly subscription charges as long as the new rate, overall, falls within the benchmarks.

47. The Act exempted a system from rate regulation if it was found to be effectively competitive.

48. The author submitted an affidavit in those proceedings. See Hazlett, 1993a.

49. Of course, there are substantial reasons to exclude A systems from the definition of "effectively competitive." First, many of them are simply systems that have failed to construct a cable plant covering an entire franchise area. Because the 30% subscribership proportion is defined as "subscribers divided by homes in franchise area," a system can qualify, even with normal penetration, just by having a sufficient number of homes in the franchise area that are not passed by cable. Second, the economics are counterintuitive: A system can be declared "effectively competitive" by having prices so high and/or service so poor that it signs up a small proportion of its potential market. This has been sarcastically designated as "the bad actor exemption." See Hazlett, 1993a.

50. Paul Fahri, "FCC Rethinks Cable TV Rules With Eye Toward Price Cuts," *Washington Post* (25 January, 1994), pp. A1, A9. After this chapter was written, the commission did recalculate the "competitive rate differential," setting it at 17% (Federal Communications Commission, 1994). (See Hazlett, 1994.)

51. This did not include à la carte or pay-per-view as noted earlier, but did include all expanded basic tiers.

52. Matt Stump, "The Big Apple Rereg Picture," *Cable World* (30 August, 1993), p. 12.

53. See Mark Robichaux, "How Cable-TV Firms Raised Rates in Wake of Law to Curb Them," *Wall Street Journal* (28 September, 1993), pp. A1, A12.

54. Vincente Pasdeloup, "More Trouble on Rereg Front: FCC, AGs investigate MSOs' new cable rates," *Cable World* (22 November, 1993), pp. 1, 65.

55. "In Rereg's Wake, C-SPAN's Losses Continue to Mount," *Cable World* (13 September, 1993), p. 10. The network had already lost 500,000 subscriber households during the summer due to "must carry" cable systems being forced to include marginally watched broadcast stations in their basic packages.

56. "Operators facing basic rate regulation Sept. 1 continue adding small, less expensive basic cable networks to system lineups. The latest beneficiary: Court TV, which says it will add 3.5 million new homes by the end of the year" (Toula Vlahou, "Regulation Bonus," *Cable World* [30 August, 1993], p. 46).

57. Rod Granger, "Will Re-regulation Give a Boost To Cable Ratings?" *Multichannel News* (25 October, 1993), p. 14.

58. The conflicting nature of the incentives facing cable system managers was described by one programming executive: "Re-regulation is like a bullet ricocheting through a room; you never know what it's going to hit" (*Ibid.*).

59. Industry jargon for worthless channels added simply to evade rate controls by diluting per-channel charges.

60. This is the ultimate in price control evasion: Exit the regulated market so as to simultaneously enter a deregulated market serving the same characteristic demand function. There are, of course, offsetting incentives that tend to discourage investment in the newly regulated sector, and the net impact on investment is ambiguous. Rate controls may create a discontinuous capital supply function, in which small increments of capital are discouraged, but large expenditures (which jump the supplier to a new technology altogether) are encouraged.

61. The race to settle land in the American West pursuant to the Homestead Act of 1863 has been characterized as a classic example of such wasteful rent seeking. Although the land that citizens homesteaded eventually became valuable, there were significant costs involved in staking claims to the land prior to the time settlement was efficient on its own terms (*i.e.*, without the added incentive provided by the competition to establish a property right). See Anderson and Hill, 1990.

62. An industry may even try to enact hostile legislation if it is helpful at the margin. Indeed, cable lobbyists actually attempted to have a reregulation bill resuscitated and passed into law in October 1990. As described in the trade press, Sen. Timothy Wirth (D-CO), a cable-friendly legislator, narrowly failed to work out a last-minute compromise with Sen. Al Gore (D-TN) after the legislation had been given up for dead. The cable industry rationale was that it was in their interests to have a weak reregulation measure pass, rather than have the issue hanging over their heads. The industry's real motive for getting legislation passed was to calm fears that the ban on telephone company competition (codified in the 1984 Act) would be removed. In September 1990, a cable industry newsletter considered the key trade-off involved in blocking reregulation legislation: "Congress is serving notice that if cable doesn't swallow its pill this year, harsher medication may be dished out next year in the form of telco entry. Rep. Ed Markey (D-MA) plans telco-cable hearings next year" Paul Kagan Associates, *SMATV News* (25 September, 1990, p. 2).

63. Richard Turner, "Sci-Fi Channel Encounters a Hard Sell Due to Competition, Reregulation Threat," *Wall Street Journal* (24 May, 1990), pp. B1, B5.

64. Edmund L. Andrews, "Cable's Big Ally on Capitol Hill: Hollywood," *New York Times* (6 January, 1992), p. D8. "Mr. Valenti [president of the MPAA] will not discuss his lobbying strategy, but he has not been shy about his distaste for the cable bill. 'We are opposed to rate regulation of our products in any form,' he said. 'That's a matter of principle.' " The trick here, of course, is that the cable bill did not attempt to control the price of movies but rather the price of movie distribution services. Normally, if distribution costs fall, demand (or imputed demand) for a product increases. The article also noted that Hollywood was disgruntled with the 1992 Cable Act due to its retransmission consent provisions, which would allow broadcasters to capture some program rents that producers, logically enough, preferred to think of as their own.

65. Vertical integration of satellite programmers is widespread in the cable television industry, and it may be that programming executives opposed rate regulation simply at the behest of corporate management (which was relatively concerned about the fortunes of its operating division). This would not explain, however, why Hollywood interests and unaffiliated programmers were equally negative about cable reregulation.

66. Paul Kagan Associates, *Cable TV Law Reporter* (30 November, 1992), p. 1.

67. See Hazlett, 1993b.

68. *Broadcasting* (1 July, 1991), "But Words Can Never Hurt You?" (Editorial), p. 78.

REFERENCES

Allard, Nicholas W. 1993a. "The 1992 Cable Act: Just the Beginning," *Hastings Comm/Ent Law Journal* 15, pp. 305–55.

___. 1993b. "Reinventing Rate Regulation," *Federal Communications Law Journal* 46 (December), pp. 63–123.

Anderson, Terry, and P. J. Hill. 1990. "The Race for Property Rights," *Journal of Law & Economics* XXXIII (April), pp. 177–98.

Andrews, Edmund. 1992. "Cable's Big Ally on Capitol Hill: Hollywood," *New York Times* (6 January), p. D8.

Averch, Harvey, and Leland L. Johnson. 1962. Behavior of the Firm Under Regulatory Constraint," *American Economic Review* (December), pp. 1052–69.

Besen, Stanley M., and Robert W. Crandall. 1981. "The Deregulation of Cable Television," *Law & Contemporary Problems* 44 (Winter), pp. 77–124.

Beutel, Philip. 1990. "City Objectives in Monopoly Franchising: The Case of Cable Television," *Applied Economics* (September), pp. 1237–47.

Broadcasting. 1991. "But Words Can Never Hurt You?" (Editorial) (1 July), p. 78.

Cable World. 1993. "In Re-reg's Wake, C-SPAN's Losses Continue to Mount" (13 September), p. 10.

Carroll, Kathleen A., and Douglas J. Lamdin. 1993. "Measuring Market Response to Regulation of the Cable TV Industry," *Journal of Regulatory Economics* 5 (December), pp. 385–99.

Cheung, Steven N. S. 1974. "A Theory of Price Control," *Journal of Law & Economics* XVII (April), pp. 53–71.

Fahri, Paul. 1994. "FCC Rethinks Cable TV Rules with Eye Toward Price Cuts," *Washington Post* (25 January, 1994) pp. A1, A9.

Federal Communications Commission [FCC]. 1993. "In the Matter of Implementation of Sections of the Cable Television Consumer Protection and Competition Act—Rate Regulation: Report and Order and Further Notice of Proposed Rulemaking," MM Docket 92-266 (Adopted 1 April, 1993; Released 3 May, 1993).

Federal Communications Commission [FCC]. 1994. "In the Matter of Implementation of Sections of the Cable Television Consumer Protection and Competition Act—Rate Regulation: Buy-Through Prohibition—Third Report and Order," MM Docket 92-266 and MM Docket 92-262 (Adopted 22 February, 1994; Released 30 March, 1994).

Federal Trade Commission [FTC]. 1990. "In the Matter of Competition, Rate Deregulation and the Commission's Policies Relating to the Provision of Cable Television Service," Before the Federal Communications Commission, MM Docket No. 89-600, Comment of the Staff of the Bureau of Economics and the San Francisco Regional Office of the Federal Trade Commission (20 April).

General Accounting Office [GAO]. 1989. *Telecommunications: National Survey of Cable Television Rates and Services* (GAO/RCED-89-193; 3 August).

———. 1990. *Telecommunications: Follow-up National Survey of Cable Television Rates and Services* (GAO/RCED-90-199; 13 June).

———. 1991. *Telecommunications: 1991 Survey of Cable Television Rates and Services* (GAO/RCED-91-195; 17 July).

Granger, Rod. 1993. "Will Re-regulation Give a Boost to Cable Ratings," *Multichannel News* (25 October), p. 14.

Hazlett, Thomas W. 1985. "The Economics of Discrimination in Rent-Controlled Housing Markets," in *Issues in Discrimination* (Washington, DC: U.S. Commission on Civil Rights).

———. 1986. "Private Monopoly and the Public Interest: An Economic Analysis of the Cable Television Franchise," *University of Pennsylvania Law Review* 134 (July), pp. 1335–1409.

———. 1990a. "Duopolistic Competition in Cable Television: Implications for Public Policy," *Yale Journal on Regulation* VII (Winter), pp. 65–119.

———. 1991. "The Demand to Regulate Franchise Monopoly: Evidence from CATV Rate Deregulation in California," *Economic Inquiry* XXIX (April), pp. 275–96.

———. 1993a. "In the Matter of Implementation of Sections of the Cable TV and Consumer Protection Act of 1992, Rate Regulation, Affidavit of Thomas W. Hazlett," MM Docket No. 92-266 (accompanying joint comments of Bell Atlantic, GTE, and Nynex; 17 June).

———. 1993b. "Cable Reregulation: The Episodes You Didn't See on C-SPAN," *Regulation* 16 (No. 2), pp. 45–52.

———. 1993c. "Why Your Cable Bill Is So High," *Wall Street Journal* (24 September), op-ed page.

Jessell, Harry A. 1992. Mooney: Rereg No Sure Thing. *Broadcasting* (4 May), p. 15.

Kimmelman, Gene. 1990. "Slam a Lid on TV Cable Rates," *Cleveland Plain Dealer* (6 June).

Leffler, Keith. 1982. "Ambiguous Changes in Product Quality," *American Economic Review* 72, pp. 956–67.

National Cable Television Association [NCTA]. 1990–94. Cable Television Developments (various issues).

National Telecommunications and Information Administration [NTIA]. 1988. "Video Program Distribution and Cable Television: Current Policy Issues and Recommendations" (Washington, DC: U.S. Department of Commerce; June).

Otsuka, Yasuji. 1993. "The Effects of Regulation on Social Welfare: Evidence from the Cable TV Industry," paper presented to the Southern Economic Association (November).

Paul Kagan Associates. 1987. *The Cable TV Financial Databook* (June).

———. 1989. *The Cable TV Financial Databook* (June).

———. 1990. *The Cable TV Financial Databook* (June).

———. 1990. *SMATV News*, (25 September).

———. 1992. *Cable TV Law Reporter* (30 November).

———. 1993. *Cable TV Investor* (12 February).

———. 1993. *Marketing New Media* (20 December).

———. 1994. "Regulating Cable Television Rates: An Economic Analysis," Working Paper No. 3, Program on Telecommunications Policy, V. C. Davis (July 1994).

Powe, Lucas A., Jr. 1987. *American Broadcasting and the First Amendment* (Berkeley: University of California Press).

Robichaux, Mark. 1993. "How Cable-TV Firms Raised Rates in Wake of Law to Curb Them," *Wall Street Journal* (28 September), pp. A1, A12.

Setzer, Florence, and Jonathan Levy. 1991. "Broadcast Television in a Multichannel Marketplace," Federal Communications Commission, OPP Working Paper No. 26 (June).

Stigler, George J. 1971. "The Theory of Economic Regulation," *Bell Journal of Economics & Management Science* 2 (Spring), pp. 3–21.

———, and Claire Friedland. 1962. "What Can Regulators Regulate? The Case of Electricity," *Journal of Law & Economics* V (October), pp. 1–16.

Stump, Matt. 1993. "The Big Apple Rereg Picture," *Cable World* (30 August), p. 12.

Turner, Richard. 1990. "Sci-Fi Channel Encounters a Hard Sell Due to Competition, Reregulation Threat," *Wall Street Journal* (24 May, 1990), pp. B1, B5.

Vlahou, Toula. 1993. "Regulation Bonus," *Cable World* (30 August), p. 46.

Zupan, Mark A. 1989. "Nonprice Concessions and the Effect of Franchise Bidding Schemes on Cable Company Costs," *Applied Economics* 21, pp. 305–23.

Part III

EMPIRICAL TRENDS
AND EVIDENCE

Chapter 8

Quality-of-Service Measurement and the Federal Communications Commission

Jonathan M. Kraushaar
FCC Common Carrier Bureau

INTRODUCTION AND OVERVIEW

Collection of quality-of-service data by regulatory agencies has historically been hampered by the difficulty in establishing uniform standards and data specifications and the cost and resources needed to collect and process the data. Allocation of resources by regulatory agencies to quality-of-service data collection has been further limited by the fact that quality-of-service data used internally by the companies is often part of a feedback mechanism within the companies and as such does not usually exhibit dramatic fluctuations requiring outside intervention. As a result of fairly stable quality-of-service levels, quality-of-service monitoring efforts at the federal level have been sporadic and are usually motivated by a significant service problem or are based on a well-defined broader objective. Significant local service problems in the late 1960s, particularly with dial-tone response, recent outage problems, and the institution of price cap regulation, have motivated the most significant regulatory responses.

This chapter describes a new quality-of-service monitoring program at the Federal Communications Commission (FCC). To place the current program into perspective one needs to consider the impact of technology and the new price cap regulatory mode now in place. The present quality-of-service monitoring program was born out of a concern that regulation under price caps might motivate the companies to place less emphasis on service quality in attempts to maximize profits. Because there was limited experience with this new form of regulation and because it was known that the companies had incentives to cut

187

costs there was a great deal of concern as to how this might affect the level of service quality of the local operating companies.

The second factor—technology—had a dual role. First, new technologies have resulted in a higher concentration of telephone traffic on a smaller number of facilities, and outages on those facilities, although infrequent, could be disastrous. This became apparent with the large AT&T switching-system failures and other significant switching failures in the operating areas of Bell Atlantic and Pacific Telesis during 1990 and 1991. Public concern had been aroused, and a clear need for regulatory intervention existed. The regulatory vehicle being established under price caps became an ideal vehicle to address these concerns at the local operating company level. The current quality-of-service monitoring program was thus initiated both to deal with broad policy objectives and to respond to specific service problems. The current effort has thus supported the work of the Network Reliability Council, which was set up to focus on reliability issues of both local and interexchange carriers.

Although technology resulted in new kinds of service quality issues, it also has provided the tools to respond more effectively. For one thing, technology has vastly reduced the cost of the data collection and analysis process. The task of reducing all the data supplied by the companies into a summary format at the FCC, for example, was accomplished by a far smaller number of personnel than would have been required prior to the age of the personal computer. Beyond this, the availability of new means to make the source data available to the public using electronic bulletin board software and personal computer technology has added a new dimension to carrier data filings that may have far-reaching implications for the regulatory process. In short, public accountability of the companies through publicly available mechanized data has not heretofore been possible. The impact of such new approaches will become apparent in the years to come.

With this backdrop, this chapter presents an initial assessment of new quality-of-service data filed with the FCC by local telephone companies. It presents an overview of the quality-of-service information now available. The source data are summarized in greater detail in an FCC report released in February 1993.[1] Due to the newness of the data and the need to establish a baseline period for evaluating any future trends, this chapter concentrates on the characteristics of the data rather than on a detailed analysis of its implications.

HISTORICAL BACKGROUND

At the end of 1983, in conjunction with AT&T's divestiture of its local operating companies, the Commission directed the Common Carrier Bureau to establish a monitoring program that would provide a basis for detecting any adverse trends in service quality. During 1985, the quality-of-service submission requirements were modified to reduce unnecessary paperwork and to ensure that the information needed by the Commission would be provided, where possible, in a more uniform format. The data were received semiannually, typically in March and August, and were the basis for FCC summary reports in June 1990 and July 1991.

With the implementation of price caps for local exchange carriers, several major changes were made beginning with reports filed in 1991. First, whereas quality-of-service reports had been received only from Bell operating companies, other companies subject to price caps were also required to submit reports on service quality. Thus, the operating companies owned by GTE, Contel, and United began to file reports. Second, quality-of-service reports were included as part of the Commission's Automated Reporting and Management Information System (ARMIS).[2] This system resulted initially from the Commission's revisions to its Uniform System of Accounts and was designed to allow financial records to be tracked for regulatory purposes in a mechanized format. The system has since been augmented to include information on telephone plant statistics associated with the local operating company infrastructure and the information on quality of service discussed here. Third, there was a considerable change in the data reported—with some items being deleted and new items added. For example, public concern over switching outages has been addressed by the program developed for price caps, and data associated with switching outages are presented in this chapter.

The data items now being monitored at the FCC resulted from a negotiation process between FCC staff and company representatives in a series of meetings hosted by the United States Telephone Association (USTA). The process assisted the companies in responding to the Commission price cap requirements in a manner that would minimize their cost by providing an opportunity to establish a common base of data already collected by the companies. In general, the collected data reflected already existing, company data collection processes, and the primary new requirement was to assemble the data in a single quarterly report on a state-by-state basis. This process has been greatly facilitated by modern computer software and hardware.

DATA CONTENT AND AVAILABILITY

As indicated earlier, the data being collected fulfill two roles. First, it addresses concerns associated with the price cap regulatory process. Second, it addresses public concern about switching outages that have become a more significant problem with larger switches and a more interconnected switching control network.

The raw quality-of-service data used in the preparation of this chapter is received by the Commission in what is called the ARMIS 43-05 filings, which are grouped into sections that are referred to as *tables*. The content and structure of this data is shown in Appendix A. Data items included in the Commission report are shown in Appendix B. Summarization of the data itself is included in Appendix C. The source data along with relevant Commission reports are available to the public on an electronic bulletin board system. The bulletin board is available 24 hours a day including weekends; however, between the hours of 8 A.M. and 1:30 P.M. usage is restricted to regulatory and governmental agencies. Each user is allotted at least 30 min daily after filling out a simple registration questionnaire.

The bulletin board operates from a standard personal computer at the Federal Communications Commission, equipped as of this writing with a 14,400-baud modem. Most of the bulletin board files are available only in a compressed format to reduce the time necessary to transmit the data, to reduce computer storage requirements, and to allow related files to be grouped into single compressed files. A special file that can be downloaded is available to decompress the files. Most files are limited in size so that any file on the board can be downloaded using a 2400-baud modem in the allotted 30-min session. Download time at the 2400-baud data rate is approximately 10 kilobytes per minute. The compressed quality-of-service files described in Appendix A and referred to as the ARMIS 43-05 reports contain the raw data from which this chapter was prepared and range in size from a few thousand bytes for companies operating in a few states to sizes somewhat exceeding 80 kilobytes for companies operating in numerous states or study areas. There is a separate file within each compressed file for each state or portion of a state, which is sometimes referred to as a study area.

When decompressed the raw data files for each study area range in size between 10 and 20 kilobytes or typically about 15,000 characters. The raw data files are in ASCII or text format but do not contain the data labels and headings necessary to identify each data item. Therefore, a special spreadsheet "template" file that can also be downloaded from the bulletin board along with the data is made available to view the data in tabular format with appropriate headings and data labels. More detail on the use of the board and the special files for decompressing and viewing the data offline are contained in information appearing on the board with each access. Instructions on the use of the board are available for downloading in another special file. Instructions can also be viewed online by examining selected bulletins. Broad public access by electronic means will provide greater public accountability to the process and should assist the regulatory process under price caps.[3]

DATA OVERVIEW AND OBSERVATIONS

Most quality-of-service data now being reported to the Commission appears in the ARMIS 43-05 report, which is filed quarterly. The ranges of selected data items associated with or calculated from these filings are graphically summarized in Figs. 8.1–8.6. These graphs were chosen to highlight the categories of data being received in connection with the ARMIS 43-05 report and to illustrate data

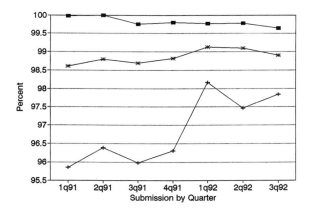

FIG. 8.1. Percentage of installation commitments met (maximum, minimum, and average reported).

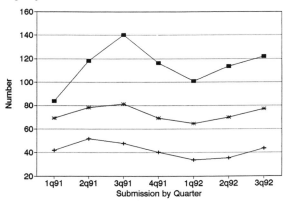

FIG. 8.2. Total trouble reports per thousand lines (maximum, minimum, and average reported).

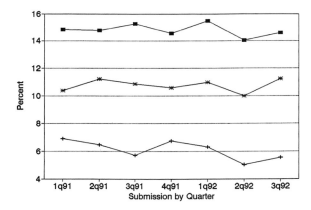

FIG. 8.3. Repeat troubles as percentage of total (maximum, minimum, and average reported).

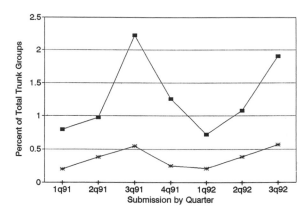

FIG. 8.4. Percentage of trunk groups over 3-month objective (maximum, minimum, and average reported).

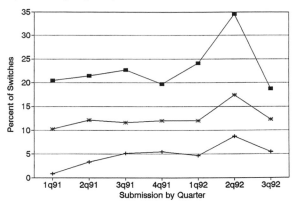

FIG. 8.5. Percentage of switches with outages (maximum, minimum, and average reported).

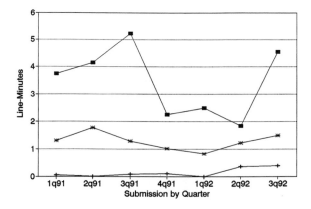

FIG. 8.6. Outage line minutes per access line (maximum, minimum, and average reported).

variability and other features of the data. The graphs include data for each major holding company (the seven Regional Bell holding companies, GTE, Contel, and the United Companies) and reflect weighted averages across individual states or study areas, along with maximums and minimums from the filed data. This type of presentation is useful in assessing the quality of the reported data and in evaluating overall trends. Given the newness of the reporting system and the sheer magnitude of the reports, there appear to be data errors, particularly in the earlier quarters. Data variability from this and other causes is demonstrated in Figs. 8.1–8.6.

The items collected by the Commission generally were selected from data sources already available within the companies and are designed to cover the major areas of quality of service affecting customers. Some items such as bit error rates on digital facilities were not included because of a lack of uniform measurements and the fact that all of the companies have not yet developed standard measurement processes.[4] Some of the items that were chosen are collected in different formats by the state regulatory commissions. Because preexisting data sources were used typically, the costs imposed on the companies consisted largely of the costs of assembling and preparing the data in a prescribed common format.

Much of the data now received, particularly the data associated with switching outages, have not been indexed or keyed to an internal company objective level. These data therefore reflect directly measured quantities. Indexed measurements or measurements tied to internal company standards are harder to standardize across companies because even if they use the same measures, the objective level on which an index is based may differ. Because indexing is used as part of well-established internal company feedback processes, index data typically exhibit small fluctuations around a well-established underlying level. Recognition of this fact and a desire to examine data closer to the measurement source led to several new data sources that were not affected by company indexing.

The overall approach was to establish a menu of quality-of-service measurements rather than trying to reduce the environment to a very small number of measurements that would tend to bias the measurements in accordance with some form of preconceived weighting scheme. The menu approach incorporating a larger number of data items is becoming more feasible because of new data processing tools and because the process was largely directed at data sources already in existence within the companies. Organization of the data along the lines of regulatory interest provides an opportunity to examine traditional areas of quality measurement from a fresh perspective utilizing a menu of data items that are available from a larger number of entities than ever before. In addition, new areas of interest relating to switching outages have been addressed.

The graphs in Figs. 8.1–8.6 highlight several normalized measurements that are of interest in a variety of areas relating to current quality-of-service measurements. Trouble reports, for example, are widely used by state regulatory agencies. Other measurements of interest depend on the perspective of the user. For example, reliability issues are addressed by the statistics such as the outage line-minutes and percentage of switches with outages that provide a measure of the impact and frequency of outages. These statistics support the work of the Network Reliability Council. The graphs illustrate the maximum, minimum, and average value associated with the filed data. Differences in trended patterns between the maximum and minimum value reported can be used to characterize the data. Typically, the maximum and/or minimum values exhibit greater variability because they characterize data outliers. This is particularly evident in the chart depicting outage line-minute data. One should also note the apparent effect of efforts to deal with problems with the data in the latest quarters shown. This is most evident in the chart depicting the percentage of installation commitments met in which the minimum value has risen sharply from the early quarters.

Rather than presenting any definitive conclusions, the aggregation and summarization of data presented here is designed to facilitate further analysis of the quality of reported data and an assessment of the program currently in place. Data summarization should enable both the Commission and the companies to improve the massive ARMIS data collection and evaluation process. Although many obvious problems have been identified and corrected, the data are subject to future updating, which hopefully will correct errors identified by this process.

One important problem relating to quality-of-service measurement in general is the continuity of measurement. Although data continuity is an important consideration, detection of errors and changes in reporting requirements that are deemed necessary may inevitably introduce discontinuities into certain data series or may eliminate those data series entirely. It is also important to note that because quality monitoring programs impose costs on the companies, historically, the data collection efforts have been vulnerable when they are perceived as outliving their usefulness. In addition, changes in technology have led to changes in the nature of the measurements required to adequately monitor service quality. Finally, the companies themselves periodically wish to change their internal measurement procedures, affecting what is reported and increasing the difficulty of long-term measurement comparisons. These factors tend to limit the number of years of data available for tracking service-quality trends. Because the present program is an offshoot of an earlier more limited one, an attempt was made to relate measurements of the two programs. Of the five areas of measurement during the period 1985–1990, only two have survived in a form that allows a longer term trend to be established: customer perception of quality-of-service levels as surveyed by the companies, and dial-tone delay. These are illustrated in Tables 8.1–8.4 in Appendix C. These items provide a very limited view of long-term trends and reflect a possible data discontinuity beginning with the new series due to known changes in the customer perception surveys and in the way the data have been developed. As presented, these data show no obvious adverse trend over the period.

Although it is premature to draw any conclusions about data trends since 1991 with only seven quarters of data, several observations can be made (some of which are summarized in the following paragraphs). It should be clearly understood, however, that at this point these observations remain tentative because the reliability of the reported data is subject to further review. Because of the relatively short time span and limited number of data points, the observations discussed here focus primarily on the typical ranges for some of the composite levels reported over the time period covered by the seven ARMIS 43-05 data measurements included in this report. This should provide some feel for the typical levels of the reported items and should assist in further understanding of the data. Many of the ranges presented here are illustrated graphically in Figs. 8.1–8.6.

The first data to consider are the number of trouble reports per thousand access lines. Nationwide, companies have typically experienced approximately 40 to 80 trouble reports per thousand access lines. The rate for residential lines is nearly twice the rate for businesses. Repeat occurrences tend to range between 5% and 15% of total trouble reports, with businesses experiencing what appears to be a slightly higher rate for repeat trouble reports than residences.

The data on switch outages indicate that nationwide, in a typical 3-month period, about 1,500 to 2,500 switching machines, representing roughly 10% to 15% of the total switches, tend to experience outages. Most of these outages last less than 2 min. The line-minutes per access line parameter was developed to compare the impact of outages lasting more than 2 min. The number of lines involved in each outage is multiplied by the outage duration and is summed over all occurrences and then is divided by the number of access lines. For example, a value of 9,000 line-minutes would be produced if a total of 9,000 lines were out of service for 1 min or if 900 lines were out of service for 10 min. The data collected thus far indicate that there have been up to 1.5 line-minutes per access line during a representative quarter. Unscheduled outage line-minutes tend to be significantly higher than the level of scheduled outage line-minutes. In addition, isolated outage levels of more than 2 line-minutes per access line have been noted in the data.

From the data one can see that installations not provided by a commitment date typically are completed up to 7 days later. However, fewer than 2% of all installations tend to be in this category. For repair of access service calls, the companies tend to respond within 5 hours for switched-access services and within 6 hours for special-access services. Response times in the 1- to 3-hour range appear frequently in the data. Data for customer complaints to regulatory agencies tend to vary widely by company. Residential complaints appear to be higher than business complaints. Finally, less than about 0.5% of the trunk groups tend to exceed the blocking objectives for the 3-month measurement period.

DATA QUALIFICATIONS AND NOTES

Although the new quality-of-service data that are being reported in the ARMIS 43-05 filings resolve many of the data concerns summarized in earlier quality-of-service reports and represent an improvement of the reporting requirements, users of these data should be aware of several pitfalls.

First, and most important, one should be aware that these data are very new. Although many problems with the data have already been identified and corrected through the many correction filings by the carriers, there are still potential flaws in the data that will only become apparent when users subject the data to further analysis or compare it to other sources. The process by which the data are checked should improve over time as the Commission and the companies progress over a normal learning curve. Although the data have been subject to an initial screening by the Commission, a number of data flaws that have not yet been corrected have been made evident by preparation of the data in this form. Holding company totals or composites and in some cases trended data items have been calculated in a consistent manner from the filed data. Some of these data items may not necessarily match company filed totals or composites. This is primarily due to different weighting methods. In addition, the carriers have updated their earlier filings numerous times. The data presented here reflect the latest updates filed with the Industry Analysis Division as of January 1993. The reader should therefore be aware that it is possible that some of the problems evident in the data presented here have already been corrected. Other problems may lead to changes in the reporting requirements themselves.

Second, although much thought has gone into the definitions of the data items, some erroneous or omitted responses have been identified. In a few instances data from subsequent quarters may reflect the correction or omission. Some of the errors may be in the process of being corrected or may not be evident until one performs further analysis with the data. Suspect data have therefore not generally been deleted or adjusted. The process of data correction should follow a normal learning curve and be resolved over time as such problems are identified and corrected. Many of the errors have been corrected by updated filings. Some of the errors have resulted from an improper reading of the instructions or a misunderstanding in the data definitions, which were worded to provide for a level of standardization without requiring costly changes to existing measurement and data collection procedures. For example, many of the companies appear to have interpreted initial trouble reports as including repeat trouble reports, even though separate categories were provided for initial and repeat trouble reports.[5]

Third, although the Commission has attempted to standardize the data requirements, one should not be lulled into the assumption that comparable data items for different companies are exactly the same. Different companies may have different procedures for collecting and presenting the data, which may affect the quality and meaning of the data provided to the Commission. Earlier quality summary reports have cautioned against direct comparisons between companies and have suggested that comparisons should be made on the basis of trends, particularly when there is little standardization in data measurement and collection procedures or when the reported data are already indexed. The current program represents a greater level of data standardization than the previous one and contains a larger number of items that are not indexed. Nonetheless, variations in the way data was collected and assembled or in the way the definitions were interpreted would tend to make comparisons of company data on the basis of trends more meaningful than single-quarter comparisons.

Finally, one should be cautious in responding too quickly to glitches or apparent sudden changes in the data, especially before getting a sense of the data. Reliability data are expected to be somewhat more erratic than the other data items. Even here, longer term patterns may be identifiable, which could assist the companies in gaining a better insight into any identified problems. Such insights should lead to more cost-effective solutions. Although the fact that the data are now being collected on a quarterly basis which permits observation of problems sooner, it also may lead an observer to draw conclusions prematurely. For example, data errors or company responses that require more than one quarter to be implemented may result in apparent abnormalities, which in fact are normal occurrences. As more experience is gained in looking at the data, one should eventually be able to recognize anomalies from normal seasonal patterns and other patterns in the data reflecting the companies' normal response in maintaining adequate service to customers. As noted in earlier quality reports, one should continue to view the data in the context of trend analysis and consider internal company response times in dealing with problems. More experience with trended data will provide a greater understanding of the subtleties inherent in the data and may eventually suggest the applicability of certain benchmarking techniques to some of the measurements.

The data items presented in this chapter are available on a study-area basis, usually on a state or a portion of a state. Further analysis supplemented with data from state regulatory commissions may be needed to address the existence of localized problems.

CONCLUSIONS

The monitoring program described in this chapter embodies several new ideas in quality-of-service monitoring, largely made possible by today's computer technology. First, rather than combining the measurements using some kind of weighting factor, the approach has been to develop a menu of items that would provide a means to gain a better understanding of service quality trends. Second, for the first time the data are available to the public in machine-readable form using electronic bulletin board technology. Third, there are extensive new data available on switching outages. It is hoped that data on small or brief outages, which are more frequent than major ones, will provide insight into the causes and most cost-effective remedies for handling the larger outages.

The current effort was designed to detect adverse quality-of-service trends under price caps and to gain a better understanding of switching outages. An understanding of the measurements and their role both by regulatory agencies and the public will contribute to the success of the approaches described here; however, the success of the program may in fact be measured by the lack of any unusual or adverse trends because it is hoped that the program itself will motivate the companies to more carefully scrutinize their quality of service. The experience gained through the current program will contribute to the future of such approaches. Finally, it should be understood that the future of the current quality-of-service program will be tied closely to the future of price cap regulation, which provided one of the key motivations for the program discussed here.

APPENDIX A: RAW DATA RECEIVED BY THE COMMISSION

The data items included in the raw ARMIS submissions by the companies are described next. These data are available in machine-readable form on the electronic bulletin board system described earlier.

A.1 Table 1

This group of data covers interexchange-switched, high-speed special, and all special-access services. Data items include:

1. Total Number of Orders or Circuits: Total installation orders or circuits for the reporting period.
2. Percentage of Commitments Met: Percentage of total installation orders met by the commitment date.
3. Average Missed Commitment in days: Average interval in calendar days between the commitment date and the day of service for all commitments not met during the reporting period.
4. Total Trouble Reports: Total number of circuit-specific trouble reports during the current reporting period.
5. Average Repair Interval: Average interval in hours to the nearest tenth from the time of the reporting carrier's receipt of the trouble report to the time of acceptance by the complaining interexchange carrier or customer.

A.2 Table 2

This group of data covers local service installations for residence and business customers subcategorized by the MSA (Metropolitan Statistical Area or area including at least one city with a population of 50,000 or an urbanized area of a population of 50,000 in an area of at least 100,000 population) and non-MSA. Data items include:

1. Installation Orders: Local Service orders or circuits.
2. Percentage of Commitments Met: Percentage of service orders completed by the commitment date.
3. Average Missed Commitment: Average interval in days from commitment date to provision of service.
4. Total Access Lines: All classifications of local-access lines including individual lines, party lines, PBX and Centrex access, coin access, foreign exchange, and WATS access.
5. Initial Trouble Reports: Complaints concerning service quality made by customers or users to local exchange carrier.
6. Repeat Trouble Reports: Trouble reports remaining unresolved within 30 days of the initial trouble report.
7. No Trouble Found: Trouble report investigation finding no discernible problem.

A.3 Table 3

These data report trunk-group blockage that prevents call completion. Data items include:

1. Total Trunk Groups: Total common trunk groups between local exchange carrier end-office and access tandem-carrying feature group B, C, or D or access traffic for which the reporting carrier is responsible.
2. Groups Measured: Common trunk groups measured during current reporting period.
3. Groups Exceeding Servicing Threshold for 3 Months: Number of common trunk groups exceeding access-tariff-measured blocking threshold (usually 2% for equal-access and 3% for non-equal-access trunks) for 3 or more consecutive months.
4. Groups Exceeding Servicing Threshold for 1 Month: Number of common trunks exceeding access-tariff-measured blocking threshold for current month.
5. Groups Exceeding Design Blocking Objectives for 3 Months: Common trunk groups exceeding equipment design blocking objectives (.5% to 1% during time-consistent busy hour of busy season) for 3 or more consecutive months.

A.4 Table 4

These data report Total Switch Downtime which includes the time when the call-processing capability for an end-office is lost, the number of incidents of less than 2 min duration, the number of switches experiencing downtime, and the number and percentage of incidents of less than 2 min duration that are not scheduled. The data are reported in the following categories:

1. Categorized by MSA and non-MSA.
2. Categorized by Switch Size: Under 1,000 lines, 1,000 to 4,999 lines, 5,000 to 9,999 lines, 10,000 to 19,999 lines, and over 20,000 lines.

Table 4a reports itemized occurrences of more than 2-min duration downtime and includes the following:

1. Explanation: Cause of downtime or scheduled or unscheduled.
2. Switch Identification: CLLI or commn language identification of switch.
3. Access Lines: Access lines served by switches and affected.
4. MSA: Y if in MSA, n if not in MSA.
5. Duration: Duration of outage in minutes to nearest tenth.

A.5 Table 5

This table reports data on Service Quality Complaints that are made to federal or state regulatory agencies categorized by MSA, non-MSA, and the total for both categories. It includes the following:

1. Business access lines in thousands.
2. Federal complaints—business users.
3. State complaints—business users.
4. Residential access lines in thousands.
5. Federal complaints—residential users.
6. State complaints—residential users.

APPENDIX B: DATA COMPONENTS INCLUDED IN THE FCC REPORT

The data summarized in the Commission report released on February 26, 1993, reflects the current emphasis on data that are closer to the measurement source. For example, rather than simply collecting data on the percentage of installations made by a commitment date, the report also reflects the number of days the company missed its commitment. These data have been derived from individual study-area data submitted by the companies by adding the numerical quantities and appropriately weighting the percentage figures. For example, the percentage of commitments met is weighted by the corresponding number of orders provided in the filed data. The summarized items included in the Commission report are as follows:

1. *Percentage of installation commitments met:* This data item provides the percentage of installations that were met by the date promised by the company

to the customer. It is shown separately for residential and business customers' local service and separately for access services provided to carriers.

2. *Average missed installation in days:* This is the average number of days beyond the commitment date that the missed installations were late. It is shown separately for access services provided to carriers and for residential and business customers' local service.

3. *Average repair interval:* This data item is the average time (in hours) for the company to repair access lines and includes subcategories for switched-access, high-speed special-access, and all special-access services. Only data for switched and special-access services provided to carriers are shown.

4. *Trouble reports per thousand access lines:* This data item is calculated as 1,000 times the sum of what was reported as "initial trouble reports" and "repeat trouble reports" divided by the number of access lines. (See endnote 5 in the text.) This item is subcategorized by MSA, non-MSA, Residence, and Business.

5. *Troubles found per thousand access lines:* This data item is calculated as described in item 4 and represents the number of trouble reports in which the company identified a problem.

6. *Repeat trouble as a percentage of trouble reports:* This data item is calculated as the number of repeat trouble reports divided by the total number of trouble reports as determined earlier. It provides a measure of the effectiveness of the company in resolving troubles at the outset. This item is subcategorized by MSA, non-MSA, Residence, and Business.

7. *Complaints per million access lines:* These data items provide the number of residential and business customer complaints per million access lines conveyed to state or federal regulatory bodies during the reporting period.

8. *Number of access lines, trunk groups and switches:* These data items provide the underlying counts of access lines in thousands, trunk groups, and switches.

9. *Switches with downtime:* This data item provides the number of switches experiencing downtime and the percentage of the total number of network switches experiencing downtime.

10. *Average switch downtime in sec per switch:* Total switch downtime divided by the total number of company switches indicates the average switch downtime in seconds per switch. It is shown for all occurrences and for occurrences greater than 2 min.

11. *Unscheduled downtime over 2 min per occurrence:* These data items provide the number of occurrences of more than 2-min duration that were unscheduled, the number of occurrences per million access lines, the average number of minutes per occurrence, the average number of lines affected per occurrence, the average number of line-minutes per occurrence in thousands, and the outage line-minutes per access line. For each outage, the number of lines affected was multiplied by the duration of the outage to provide the "line-minutes" of outage. The resulting sum of these represents the total outage line-minutes. This number

was divided by the total number of access lines to provide the line-minutes per access line and by the number of occurrences to provide the line-minutes per occurrence. This categorizes the normalized magnitude of the outage in two ways and provides a more realistic means to compare the impact of such outages between companies. A separate table is provided for each company showing the number of outages and outage line-minutes by cause.

12. *Scheduled downtime over 2 min per occurrence:* This item is identical to item 11, except it consists of scheduled occurrences rather than unscheduled occurrences.

13. *Trunk groups with blocking over 3-month objective as a percentage of total trunk groups:* This data item provides the percentage of trunk groups exceeding the objective for blocking for 3 consecutive months.

APPENDIX C: DATA SUMMARIZED
IN THE FCC REPORT

Tables 8.1 through 8.4, included in this Appendix, summarize data received since 1985. Table 8.5 is an example of the data presented in the recent quality-of-service summary issued by the Commission. A similar presentation for the Bell operating companies, the GTE companies, the CONTEL companies, and the UNITED companies is presented in the Commission report. Data on dial-tone response filed since 1985 now appear in the ARMIS 43-06 filing. Paper copies of the customer perception survey data are still filed, but these data are not contained in the mechanized ARMIS reporting formats.

The impact of new technology is reducing the significance of some of the measurements filed since 1985. For example, the dial-tone delay measurement is becoming less useful with the increasing number of digital switches, in which service is unlikely to be affected by slowed dial-tone response.

The all-company composites shown in Tables 8.1 through 8.4 are calculated in a manner consistent with earlier reports as the unweighted average of the available data compiled for the individual Bell Holding Companies. One should note that data for 1991 and 1992 may differ from the earlier part of the series. Such discontinuity is due to changes in reporting procedures. Bell Atlantic has reported changes to its customer perception surveys, which are being reflected in post-1990 data and may have resulted in data discontinuities. Other companies, including NYNEX and Pacific Telesis, have indicated that they have made or are planning similar changes.

Tables 8.1 through 8.3 cover customer satisfaction surveys performed by the companies. Table 8.4 shows the percentage of offices providing less than a 3-sec dial-tone delay. Transmission quality data have not been included in this report as they do not cover transmission quality on the increasing number of digital transmission facilities that presently comprise over 95% of the interoffice facility

TABLE 8.1
Percentage of Customers Satisfied—Residential

Company	1985	1986		1987		1988		1989		1990		1991		1992
		1H	2H	1H	2H	1H	2H	1H	2H	1H	2H	1H	2H	1H
AMERITECH	92.8	94.8	94.0	94.7	94.1	95.0	94.2	94.8	93.6	94.4	94.3	95.3	94.9	95.4
BELL ATLANTIC	92.4	93.4	93.0	94.4	90.2	92.3	92.1	91.8	93.3	94.6	93.9	95.6	95.7	94.9
BELLSOUTH	92.0	92.8	92.8	94.2	94.0	93.9	93.6	94.1	93.2	94.9	94.9	95.5	NA	92.7
NYNEX	93.7	93.5	92.9	93.6	93.6	94.5	94.0	94.2	94.1	92.8	93.7	94.7	93.6	92.6
PACIFIC TELESIS	94.1	93.0	94.4	95.6	96.1	95.8	95.7	96.9	96.0	96.5	95.5	96.7	96.7	95.5
SOUTHWESTERN	97.6	97.6	95.5	96.1	95.8	96.3	96.3	96.5	96.4	96.8	96.6	96.8	96.5	96.6
U S WEST	91.9	92.7	92.4	93.3	94.1	93.3	93.3	92.1	91.4	91.8	91.2	93.6	93.1	92.4
COMPOSITE	93.5	94.0	93.6	94.5	94.0	94.4	94.2	94.3	94.0	94.5	94.3	95.5	95.1	94.3

Note. Holding company data in this table are derived as an unweighted average of available operating company results. Composites are unweighted averages of holding companies. Please refer to text for accompanying notes and data qualifications.[1]

TABLE 8.2
Percentage of Customers Satisfied—Small Business

Company	1985	1986 1H	1986 2H	1987 1H	1987 2H	1988 1H	1988 2H	1989 1H	1989 2H	1990 1H	1990 2H	1991 1H	1991 2H	1992 1H
AMERITECH	90.6	93.8	93.8	94.4	94.4	94.6	93.9	94.6	94.0	94.6	94.9	95.7	95.4	95.8
BELL ATLANTIC	89.9	91.9	91.7	93.3	90.7	92.3	92.0	NA	NA	NA	NA	94.9	95.1	93.8
BELLSOUTH	92.0	93.3	93.3	94.5	94.5	95.0	94.8	94.7	94.7	95.2	95.7	94.9	NA	94.5
NYNEX	91.6	91.6	91.2	92.3	92.2	93.9	93.4	93.7	93.5	91.9	92.7	93.9	92.9	92.2
PACIFIC TELESIS	94.2	91.7	93.4	94.5	94.0	93.9	94.1	95.6	95.3	95.9	94.9	96.1	96.1	94.0
SOUTHWESTERN	97.1	97.0	94.6	95.0	95.0	95.8	95.6	95.8	95.5	95.9	95.7	96.4	96.2	96.4
U S WEST	89.4	91.1	91.1	92.1	93.5	92.6	92.4	90.4	89.8	90.7	89.8	92.1	90.7	92.2
COMPOSITE	92.1	92.9	92.7	93.7	93.5	94.0	93.7	94.1	93.8	94.0	94.0	94.9	94.4	94.1

Note. Holding company data in this table are derived as an unweighted average of available operating company results. Composites are unweighted averages of holding companies. Please refer to text for accompanying notes and data qualifications.[1]

TABLE 8.3
Percentage of Customers Satisfied—Large Business

Company	1985	1986		1987		1988		1989		1990		1991		1992
	1H	1H	2H	1H	2H	1H	2H	1H	2H	1H	2H	1H	2H	1H
AMERITECH	89.1	90.7	90.2	90.0	91.4	95.1	93.6	93.9	94.7	94.7	95.1	95.9	96.2	96.2
BELL ATLANTIC	93.1	93.3	94.0	94.0	95.0	96.0	95.7	98.0	96.0	97.3	97.0	97.6	97.1	98.2
BELLSOUTH	89.9	94.2	94.2	95.0	94.9	95.4	93.9	93.9	94.1	94.6	94.6	95.8	NA	94.8
NYNEX	94.8	96.5	97.0	91.5	91.6	93.3	92.0	94.0	93.5	93.5	93.2	94.2	94.1	90.9
PACIFIC TELESIS	90.4	92.0	95.9	94.3	93.3	92.7	94.7	95.0	95.0	93.0	94.0	94.3	94.3	90.0
SOUTHWESTERN	91.3	91.4	92.3	93.9	94.4	95.4	95.4	94.3	94.0	94.6	95.3	97.4	97.3	96.6
U S WEST	92.2	NA	95.1	NA	96.3	NA	95.5	92.1	89.0	91.1	92.4	NA	NA	NA
COMPOSITE	91.5	93.0	94.1	93.1	93.8	94.6	94.4	94.5	93.8	94.1	94.5	95.9	95.8	94.5

Note. Holding company data in this table are derived as an unweighted average of available operating company results. Composites are unweighted averages of holding companies. Please refer to text for accompanying notes and data qualifications.[1]

TABLE 8.4
Percentage of Offices Providing Dial Tone in Less Than 3 Sec

Company	1985	1986		1987		1988		1989		1990		1991		1992
		1H	2H	1H	2H	1H	2H	1H	2H	1H	2H	1H	2H	1H
AMERITECH	98.2	98.3	98.6	98.6	99.1	99.0	99.6	99.4	99.0	98.3	98.2	99.4	98.8	99.5
BELL ATLANTIC	97.8	98.2	98.6	97.8	98.8	99.0	99.3	99.3	99.1	98.4	99.2	99.5	99.6	99.8
BELLSOUTH	96.8	96.3	96.3	95.0	96.0	97.4	97.6	97.8	98.2	98.4	98.0	99.2	99.2	99.3
NYNEX	96.6	98.5	99.7	99.8	99.6	99.7	99.7	99.8	99.8	99.5	99.6	100.0	99.6	99.8
PACIFIC TELESIS	100.0	99.9	100.0	99.7	99.7	99.7	99.7	99.7	99.1	99.7	99.6	99.7	99.7	100.0
SOUTHWESTERN	97.9	98.3	97.9	98.4	98.1	99.3	99.4	99.3	99.4	99.2	99.3	97.8	97.7	98.1
U S WEST	96.7	97.8	97.2	98.2	98.4	98.8	99.1	98.9	99.4	99.0	98.9	99.3	99.2	99.6
COMPOSITE	97.7	98.2	98.3	98.2	98.5	99.0	99.2	99.2	99.1	98.9	99.0	99.3	99.1	99.4

Note. Holding company data in this table are derived as an unweighted average of available operating company results. Composites are unweighted averages of holding companies. Please refer to text for accompanying notes and data qualifications.[1]

TABLE 8.5a

Ameritech—Installation, Maintenance, and Customer Complaints

	Reporting Period						
	1Q'91	2Q'91	3Q'91	4Q'91	1Q'92	2Q'92	3Q'92
ACCESS SERVICES PROVIDED TO CARRIERS—SWITCHED ACCESS							
Percentage of Installation Commitments Met	99.5%	99.9%	100.0%	99.5%	99.3%	98.6%	99.5%
Average Missed Installation (days)	NA	0.7	0.4	0.8	3.3	1.9	4.2
Average Repair Interval (hours)	2.3	2.6	2.5	1.8	1.6	1.5	1.6
ACCESS SERVICES PROVIDED TO CARRIERS—SPECIAL ACCESS							
Percentage of Installation Commitments Met	99.8%	99.9%	99.4%	99.8%	99.8%	99.9%	99.8%
Average Missed Installation (days)	0.1	1.6	1.8	5.2	3.0	3.4	5.0
Average Repair Interval (hours)	2.2	2.4	2.4	2.2	2.2	2.2	2.3
LOCAL SERVICES PROVIDED TO RESIDENTIAL AND BUSINESS CUSTOMERS							
Percentage of Installation Commitments Met	99.6%	99.6%	99.5%	99.6%	99.7%	99.7%	99.6%
Residence	99.6%	99.6%	99.6%	99.6%	99.7%	99.7%	99.7%
Business	99.4%	99.4%	99.3%	99.5%	99.4%	99.4%	99.3%
Average Missed Installation (days)	2.8	2.8	2.8	3.0	3.0	2.7	3.2
Residence	2.7	2.5	2.8	2.6	2.8	2.5	2.7
Business	3.4	3.3	2.8	3.6	2.4	2.2	2.8
Trouble Reports per Thousand Lines	68.1	85.7	78.0	69.1	54.6	79.4	71.6
Total MSA	NA	NA	NA	NA	66.4	78.7	70.9
Total non-MSA	NA	NA	NA	NA	64.2	86.1	78.8
Total Residence	97.8	123.6	113.0	100.1	78.3	96.7	89.1
Total Business	7.5	8.5	7.8	7.4	40.9	43.1	35.4
Troubles Found per Thousand Lines	51.0	66.0	59.2	52.1	34.7	58.1	45.2
Repeat Troubles as a Pct. of Trouble Reports	10.2%	10.1%	9.8%	10.1%	11.5%	9.5%	14.6%
Total Residence	10.3%	10.1%	9.8%	10.1%	9.8%	9.7%	15.0%
Total Business	9.7%	10.0%	10.0%	9.8%	8.3%	8.5%	12.7%
Customer Complaints per Million Access Lines							
Residential	4.2	5.0	2.9	2.3	2.8	2.6	5.2
Business	1,001.7	712.2	2.2	1.1	0.9	0.7	1.7

Note. Please refer to text for notes and data qualifications.[1]

TABLE 8.5b
Ameritech—Switch Downtime and Trunk Blocking

	Reporting Period						
	1Q'91	2Q'91	3Q'91	4Q'91	1Q'92	2Q'92	3Q'92
Total Access Lines in Thousands	16,586	16,584	16,772	16,825	16,634	16,658	16,780
Total Trunk Groups	1,207	1,176	1,172	1,146	1,153	1,146	1,143
Total Switches	1,396	1,368	1,384	1,420	1,422	1,440	1,443
Switches with Downtime							
Number of Switches	41	45	105	245	138	205	271
As a Percentage of Total Switches	2.9%	3.3%	7.6%	17.3%	9.7%	14.2%	18.8%
Average Switch Downtime in sec per Switch							
For All Occurrences	5.9	22.6	28.1	56.0	63.0	66.1	204.2
For Unscheduled Occurrences More than 2 min.	1.5	17.4	13.4	38.3	55.6	50.3	173.3
For Unscheduled Downtime More than 2 min.							
Number of Occurrences	11	20	43	28	37	44	68
Occurrences per Million Access Lines	0.66	1.21	2.56	1.66	2.22	2.64	4.05
Average Outage Duration in Minutes	3.2	19.8	7.2	32.4	35.6	27.5	61.3
Average Lines Affected per Occurrence in Thousands	25.5	35.1	24.1	13.8	7.0	7.2	14.2
Outage Line Minutes per Occurrence in Thousands	75	756	149	275	332	156	1,122
Outage Line Minutes per Thousand Access Lines	50	912	381	458	739	412	4,546
For Scheduled Downtime More than 2 Min							
Number of Occurrences	32	38	84	63	32	71	118
Occurrences per Million Access Lines	1.93	2.29	5.01	3.74	1.92	4.26	7.03
Average Outage Duration in Minutes	3.2	3.2	4.0	3.6	3.0	3.4	4.3
Average Lines Affected per Occurrence in Thousands	38.6	25.2	19.6	19.5	21.1	20.5	10.9
Outage Line Minutes per Occurrence in Thousands	93	80	74	70	57	63	47
Outage Line Minutes per Thousand Access Lines	180	184	369	262	109	269	328
Pct. Trunk Grps. Exceeding Blocking Obj. 3 Months	0.66%	0.51%	0.68%	0.09%	0.17%	0.26%	0.35%

Note. Please refer to text for notes and data qualifications.[1]

TABLE 8.5c
Ameritech—Switch Downtime Causes

	Reporting Period						
	1Q'91	2Q'91	3Q'91	4Q'91	1Q'92	2Q'92	3Q'92
TOTAL NUMBER OF OUTAGES							
1. Scheduled	NA	NA	NA	NA	32	71	118
2. Procedural Errors—Telco. (Install./Maint.)	NA	NA	NA	NA	2	1	1
3. Procedural Errors—Telco. (Other)	NA	NA	NA	NA	3	0	2
4. Procedural Errors—System Vendors	NA	NA	NA	NA	2	1	4
5. Procedural Errors—Other Vendors	NA	NA	NA	NA	1	1	1
6. Software Design	NA	NA	NA	NA	2	4	8
7. Hardware Design	NA	NA	NA	NA	1	1	0
8. Hardware Failure	NA	NA	NA	NA	7	12	33
9. Acts of G-d	NA	NA	NA	NA	4	1	0
10. Traffic Overload	NA	NA	NA	NA	0	0	1
11. Environmental	NA	NA	NA	NA	0	1	0
12. External Power Failure	NA	NA	NA	NA	0	0	0
13. Massive Line Outage	NA	NA	NA	NA	0	0	0
14. Remote	NA	NA	NA	NA	12	13	13
15. Other/Unknown	NA	NA	NA	NA	3	9	5

(Continued)

TABLE 8.5c
(Continued)

				Reporting Period			
	1Q'91	2Q'91	3Q'91	4Q'91	1Q'92	2Q'92	3Q'92
TOTAL OUTAGE LINE MINUTES PER THOUSAND ACCESS LINES							
1. Scheduled	NA	NA	NA	NA	109.3	269.3	327.8
2. Procedural Errors—Telco. (Install./Maint.)	NA	NA	NA	NA	5.2	2.6	10.4
3. Procedural Errors—Telco. (Other)	NA	NA	NA	NA	8.0	0.0	543.8
4. Procedural Errors—System Vendors	NA	NA	NA	NA	216.6	4.5	21.0
5. Procedural Errors—Other Vendors	NA	NA	NA	NA	1.6	1.7	1.2
6. Software Design	NA	NA	NA	NA	1.6	3.6	246.5
7. Hardware Design	NA	NA	NA	NA	0.3	13.3	0.0
8. Hardware Failure	NA	NA	NA	NA	168.8	320.6	3,639.1
9. Acts of G-d	NA	NA	NA	NA	26.2	3.0	0.0
10. Traffic Overload	NA	NA	NA	NA	0.0	0.0	43.4
11. Environmental	NA	NA	NA	NA	0.0	2.4	0.0
12. External Power Failure	NA	NA	NA	NA	0.0	0.0	0.0
13. Massive Line Outage	NA	NA	NA	NA	0.0	0.0	0.0
14. Remote	NA	NA	NA	NA	310.0	43.8	33.9
15. Other/Unknown	NA	NA	NA	NA	1.1	16.7	6.4

Note. Please refer to text for notes and data qualifications.[1]

210

mileage as reported to the Commission by the companies. Furthermore, these data exhibited a larger data discontinuity from the earlier data series than the data shown in Tables 8.1–8.3. This appears to have resulted from changes in reporting procedures and data formats. Data on blocking and on-time installations have been modified considerably and are not comparable to the prior data series.

ACKNOWLEDGMENT

The views expressed are those of the author and do not necessarily reflect the views of the Federal Communications Commission.

ENDNOTES

1. Kraushaar, J., *Quality of Service for the Local Operating Companies Aggregated to the Holding Company Level*, Common Carrier Bureau, Federal Communications Commission, Washington, DC, February 1993.
2. The ARMIS database includes a variety of financial and infrastructure company-mechanized reports in addition to the quality-of-service reports. Data are available disaggregated to a study area or state level.
3. Individuals wishing to access the bulletin board may do so by dialing (202) 418-0241 from an appropriately equipped personal computer.
4. The Exchange Carriers Standards Association T1Q1.4 Committee is addressing performance limits for digital-transmission quality parameters.
5. The companies apparently count all trouble reports associated with a single unresolved trouble as a single initial trouble. A trouble recurring after it is initially resolved or after a specified time is reported as a repeat trouble, but many of the companies also count this as a new initial trouble. Due to a misunderstanding associated with this, at least some of the trouble report data in the Commission report in effect may reflect a double counting of repeat trouble reports. Although errors associated with this misunderstanding are also reflected in the figures shown in this chapter, the data fluctuations displayed shown would still properly reflect any trend or lack thereof because consistent procedures were used to assemble and format the data. This and other issues involving the data, the definitions, and improvements to the process are part of the Commission's ongoing evaluation.

Chapter 9

The Impact of Local Competition on Network Quality

Richard G. Tomlinson
Connecticut Research

1. INTRODUCTION

Quality is a moving target. When the role of the telecommunications network was to transport analog voice signals, quality was a relatively simple issue. Good quality meant affirmative responses to the questions: (a) When I pick up the telephone do I usually get dial tone? (b) Does the call go through promptly? or (c) When I reach the called party can the conversation be clearly heard? These subjective standards could be passed even under circumstances in which quantitative technical measurements would have shown that significant signal distortion, delay, and noise interference were present.

As the strategic importance of business communications increased and as networks evolved to incorporate digital data and multiplexed voice transport, more stringent, quantitative quality standards became necessary. If a digital line carrying hundreds of multiplexed calls is out of service even briefly or if a line carrying critical data experiences bit errors due to a high noise environment, significant financial losses can result.

In the mid- to late 1980s, small, entrepreneurial, Alternate Local Transport (ALT) companies formed to pursue the niche market for high-reliability and high-quality, dedicated, digital transport. Using fiber optics, state-of-the-art electronics, and redundant network elements, they were able to offer high-speed digital circuits of greater quality than previously available in the public network. Even though these fiber networks were of limited extent, sometimes spanning only a few miles in and around an urban center, they served important markets.

They provided long-distance carriers (IXCs—Interexchange Carriers) with local links between their multiple-switch locations (POP—Point of Presence) and/or to the switches of other IXCs with nearby POPs.

Beyond serving the IXC market, the ALT companies were also able to address a niche market of select corporate end-users. Large corporations that had locations on the ALT network could be provided with high-speed, dedicated access linking their PBXs directly to long-distance carrier POPs. Such dedicated circuits, also available from the Local Exchange Carriers (LECs), are known as *special-access* circuits, and hence ALTs are also called Competitive Access Providers (CAPs). The presence of this ALT/CAP competition stimulated a major LEC competitive response involving network development, service, and pricing for dedicated transport.

Although dedicated transport was the point of entry, competitive activity in the local exchange is growing rapidly both in terms of services offered and areas served and is commonly believed to be leading the market toward a universally competitive telecommunications marketplace in the United States, including local as well as long-distance service. New participants (e.g., cable television companies, out-of-region LECs, PCS companies, etc.) are taking an interest in this competition and are investing in alternatives to the public network. Regulatory developments at both the FCC and state level are progressively increasing the arena in which competitive entry is possible. Interconnection between ALT networks and LEC networks for both dedicated and switched access has been mandated by the FCC for interstate traffic and by some PUCs for intrastate traffic. We are headed toward a "network of networks"—an infrastructure of interconnected but competing networks of varying quality. In this future environment the potential results of competition on network quality will be more complex and perhaps more controversial.

The purpose of this chapter is to address both the current and future impact of this competition on network quality.

2. LOCAL TRANSPORT FOR INTEREXCHANGE CARRIERS

Most ALT companies began operations by serving as interexchange carriers. One application was to provide trunking between an IXC's multiple switches or "points of presence" in a large city. Another was to provide trunks between the "points of presence" of different IXCs so that they could aggregate traffic, lease capacity, and so on. Traditionally, the IXCs' choices had been either to build these facilities or to buy dedicated circuits from the LEC. The presence of an ALT created a third choice. The advantages for the interexchange carrier were that the new ALT network provided service that incorporated the latest fiber and digital technology, was priced under the LEC price umbrella, could be used to diversify network structure, was more customer responsive, and created pressure on the LEC to lower prices, improve service, and improve network quality. Thus, IXCs saw ALTs as a means to increase the quality of their local-access facilities while reducing their costs.

The IXCs were interested in obtaining local, high-speed digital links at DS-1 and DS-3 rates (1.544 Mbps and 45 Mbps—capable of supporting 24 and 672 voice channels, respectively). As greater numbers of voice-grade circuits are multiplexed onto a single, high-speed, digital-fiber channel, the potential outage impact produced by a fiber-cable cut or other malfunction is multiplied, increasing the need for high reliability. Initial ALT service and network quality were strongly influenced by the IXCs. As the dominant, and sometimes sole, customer, the major IXCs imposed their own certification standards as a prerequisite for doing business with an ALT.

High reliability was achieved by using multiplexers with redundant electronics and automatic switchover in case of component failure. Further reliability was achieved by adopting robust network architectures with the capability to automatically switch at high speed (~50 msec) from transmitting over a primary fiber to an alternate fiber in case of the loss of signal in the primary fiber. Such fiber circuits were called *self-healing* and produced network availability numbers that were previously unavailable. One measure of network reliability is *circuit availability*, presented in the form of the percentage of the time the average circuit was available for service during the year. In 1988, Teleport Communications Group, then serving predominantly the New York IXC market, reported[1] that for the year it achieved an average circuit availability of 99.99%, which is equivalent to 52.6 min of outage per year. No copper-based circuits could guarantee such performance.

Techniques for achieving high reliability continued to advance, with secondary fiber paths being physically separated from primary fiber paths to decrease vulnerability to a common disaster. Network monitoring was used to detect and counteract system quality degradation before hard failure had occurred. Diverse fiber routing usually took the form of fiber *rings* in which the primary and secondary fiber paths operated in counterrotation around the ring so that no two points could be isolated by a single cable cut. By 1991, Teleport Communications Group was reporting average circuit availability of 99.999% (equivalent to 5.26 min of outage per year) in its new Boston network. (The achievement of "five nines" is now a performance requirement for many ALT local operations managers.)

Since the initiation of fiber-optic-based competition in local transport in 1985, such competitive activity has grown steadily. By the end of 1993, over 30 ALT/CAP companies were providing services on networks in 72 Metropolitan Statistical Areas.[2] Although these networks covered over 5,000 miles, linked more than 4,000 major commercial buildings, and provided access to hundreds of long-distance carrier "points of presence," total ALT industry revenues for

1993 were less than $350 million (0.4% of total LEC revenues) with only $200 million due to dedicated transport. However, this relatively modest market capture by ALT/CAPs stimulated strong reductions in LEC tariffed prices for dedicated transport, which declined by more than 50% over this period. Interexchange carriers saw even larger price reductions on their high-traffic density routes. Through combined volume and term discounts, additional savings of up to 70% of tariffed prices were attainable.[3]

3. SPECIAL-ACCESS TRANSPORT

Nearly all ALT/CAP companies have seen their business mix begin with an initial dependence on providing IXCs with POP-to-POP links and shift to a broader market, providing end-users to POP special-access links. This progression is illustrated in the case of Intermedia Communications of Florida (one of the few ALTs whose stock is publicly traded) as shown in Table 9.1.[4]

Dedicated special-access fiber links (usually at DS-1 speed) between the premises of large telecommunications end-users and the POPs of their chosen IXCs reduce the cost of access because the circuit price is flat rated and not a function of traffic volume. Most special-access end-users are in telecommunications-sensitive industries such as financial services, telemarketing, and so on.

Although their influence is not as great in the special-access market, IXCs still play a strong role. Traditionally, IXCs have often been the purchasers of special access from LECs on behalf of end-users. Even when end-users deal directly with LECs and ALTs, their choice of vendor can depend on IXC recommendations. Therefore, the IXCs' opinion can also drive vendor network quality in the special-access market.

The market demand for ALT/CAP special-access service suddenly accelerated following a fire on May 8, 1988, in a Hinsdale, IL wire center of the Illinois Bell Telephone company. Most of the 42,000 local lines and 118,000 local and long-distance trunks in the wire center were put out of service,[5] impacting thousands of business and residential customers in Chicago's southwestern sub-

TABLE 9.1
Intermedia Communications—Evolution to End-User Dependence

Year	% Recurring Revenue From End-User to POP Transport
1988	0%
1989	28%
1990	38%
1991	42%
1992	48%
1993	53% (Est.)

urbs. The last circuit was not restored for 28 days. Overnight, redundancy and route diversity for special-access circuits became a focused concern of telecommunications managers at large government and corporate installations throughout the country.

Telecommunications managers sought to diversify their circuits among multiple carriers rather than to totally displace any one carrier. There is evidence that end-users with critical telecommunications requirements view the ability to acquire access circuits from multiple vendors as desirable no matter how high the quality of the network offered by any single vendor. In a section of the Boston financial district where MFS, Teleport, and New England Telephone all serve the same buildings with fiber circuits, a survey of 21 major end-users was taken by Connecticut Research in 1990.[6] The survey showed that 24% of the sample had special-access circuits from two vendors, and another 24% had such circuits from all three vendors. Although there was some functionality differentiation among the vendor services, the primary driver for the end-users seemed to be to obtain the maximum possible diversity in both network and service provider in order to assure maximum network reliability and survivability.

Local Exchange Carrier central office wire centers are often points of concentration for dedicated as well as switched traffic. In addition to a central office switch and other equipment for handling switched traffic, some wire centers are nodes for dedicated transport and contain equipment to multiplex, demultiplex, and cross-connect nonswitched traffic. In September 1992, the FCC mandated[7] that LECs must file interconnection tariffs so that ALT/CAP companies could interconnect with dedicated-access circuits at the LEC central office wire center. Except in the case of space limitations or existing PUC regulations, this interconnection was to be accomplished through physical collocation of ALT/CAP equipment in the LEC wire center. Otherwise, the interconnection was to be accomplished by virtual collocation at a point near the wire center under conditions functionally equivalent to physical collocation.

This greatly expanded the ALT/CAP-addressable market because off-net locations could be served. End-users could buy dedicated LEC transport from their location to the wire center and transfer to the ALT/CAP network for dedicated transport from the wire center to the IXC POP. This interconnection potentially raised a new set of issues bearing on network quality because overall network performance now required a certain degree of operational cooperation between competitors.

These operational issues were largely solved without major problems. LECs built out cages in the wire center for interconnectors (including ALT/CAP companies) who paid an installation charge and a monthly lease based on floor space. Interconnectors placed their own equipment inside the locked cage, purchased electrical power from the LEC, and extended their network management systems to include performance surveillance over the link into the wire center.

4. SWITCHED-ACCESS TRANSPORT

The FCC extended interconnection to include switched-access traffic as well as dedicated-access traffic. Tier 1 (>$100 M annual revenue) LECs were required to file tariffs by November 1993. The transport portion of switched access (from the central office to the IXC POP) appears no different than dedicated access. In fact, LECs have been able to increase the efficient use of dedicated circuits by combing switched- and dedicated-access traffic to use the available bandwidth (a process called *ratcheting*) of the same trunk. Switched-access traffic differs from dedicated-access traffic in that it enters the central office on a switched circuit and is billed on the basis of multiple elements including minutes of use. With collocation, ALT/CAP companies can compete only for the transport element of switched access.

Because switched-access traffic greatly exceeds dedicated-access traffic, regulators were concerned that the financial impact on LECs not be too abrupt and imposed a residual interconnection charge (RIC), initially at 80%, on interconnectors. Interconnection for switched access, as well as dedicated access, potentially increases the number of economically desirable points of interconnection from a few high-density central office wire centers in urban areas to many wire centers, even in smaller cities. Therefore, future issues of network quality will increasingly take place in an environment of a web of intimately interconnected, multiple competing networks.

5. LOCAL EXCHANGE CARRIER RESPONSE TO COMPETITION

Local exchange carriers responded to the competitive challenge to meet the high reliability needs of IXCs and special-access end-users and installed their own self-healing and diverse fiber rings. Truly diversified fiber routing throughout an entire fiber system has been difficult for both LECs and ALTs to achieve; and, in practice, many systems retain some "spurs" that are subject to single-point failure. Most LEC conduit was originally installed for star or tree-and-branch cable deployment and not for rings. Both LECs and ALTs find that building owners often object to the construction of additional telecommunications entrance facilities through the walls of their structures.

In their reports of fiber deployment, the Industry Analysis Division of the Common Carrier Bureau of the FCC began including information on LEC fiber rings in 1989.[8] Many of these LEC installations are not counterrotating rings but are path-switched, multiple-fiber systems that are often physical stars but logical rings. Generically, all these self-healing networks are referred to as *rings*. The 1989 FCC report showed that LECs had deployed fiber rings in 14 cities, primarily in the vicinity of competing ALT networks, and the 1990 report[9] showed 56 cities. By 1991,[10] the number of cities had grown to 127, and by 1992[11] to 188. GTE alone began a 50-city deployment of advanced fiber rings in 1993.

US West announced in May 1990 that it would deploy fiber rings in five major cities—Denver, Minneapolis/St. Paul, Seattle, Portland, and Phoenix—with circuit availability performance of 99.99%. This announcement also included the most aggressive performance-guarantee standard publicly offered. US West guaranteed that any customer on the ring that suffered a network outage in excess of 1 sec would receive a full month's credit of the circuit's lease rate. This was followed by other LECs offering performance guarantees for transport on new self-healing fiber networks.

In addition to deployment of self-healing fiber rings in high-density traffic areas, LECs created special units such as NYNEX's "Enterprise Services" to provide high-quality, expedited service to the customers served by these networks.

6. NETWORK QUALITY RESULTS

The Regional Bell Operating Companies have published[12] their standards for network availability of dedicated access circuits as shown in Table 9.2.

These are operational standards, and actual network availability achieved has not been reported. Because the network technology and equipment used by RBOCs and by ALT/CAP competitor networks is equally available from the same vendors, new fiber-network installations should be capable of identical system performance. However, RBOCs and other established carriers must deal with the fact that they have an installed base of copper and older fiber circuits as well as the more capable new fiber circuits.

Bell Communications Research (Bellcore), the central research organization of the RBOCs, has published standards[13] for fiber-network performance. This reference establishes an outage standard for short (<25 mile) interoffice fiber trunks[14] of 16 min per year (99.997% availability) for DS-1 and 8 min per year (99.9985% availability) for DS-3 transport.

TABLE 9.2
RBOC Published Standards for Network Availability—1992

Carrier	Network Availability Standard
Ameritech	99.975%
Bell Atlantic	99.925%
NYNEX	99.700% (IntraLATA)
NYNEX	99.925% (InterLATA)
Pacific Bell	99.975%
Southwestern Bell	99.975%
US West	99.700% (99.990% Fiber Ring)

Metropolitan Fiber Systems (MFS), one of the largest ALTs with 14 networks, has been outspoken on the issue of network quality. They have noted that network quality is more than physical parameters such as network circuit availability. Network quality from the end-user's perspective also includes the organizational responsiveness of the carrier in terms of installation and repair intervals. Therefore, MFS has published[15] comparisons of its own standards and performance for network availability, installation interval, and service repair interval versus those of the Regional Bell Operating Companies (RBOCs).

MFS indicated that, although its standard for network availability was 99.99%, it routinely exceeds this standard. Its average circuit availability in the first quarter of 1992 was 99.99898% (5.36 min per year) for DS-1 circuits and 99.99976% (1.26 min per year) for DS-3 circuits. For this same period, MFS achieved average installation intervals of 7.8 calendar days and 7.6 calendar days, respectively, for DS-1 and DS-3 circuits. MFS also achieved average repair intervals of 90 min and 23 min, respectively, for DS-1 and DS-3 circuits. These intervals are significantly better than the published standards of the LECs.

The only relevant parameters routinely reported through the FCC's Automated Reporting and Management System (ARMIS) for the LECs is Average Repair Interval and the Average Missed Installation Days for special-access services. Table 9.3 provides a sampling of these data.[16]

When Teleport began operations, installation intervals for DS-1 circuits from New York Telephone exceeded 90 days, and DS-3 circuits were not available. Under the competitive pressure from Teleport, NYT installation intervals have steadily decreased. Noting that its market share for private lines in Manhattan had fallen below 65%, NYT launched a "Take Back New York City" campaign in April 1993. A key element of this campaign is the NYNEX Enterprise Services offering. This service, initially available only in lower and midtown Manhattan, offers 10 speeds (including DS-1 & DS-3) of up to 100 Mbps. The network

TABLE 9.3
RBOC Reported Performance for Installation and Repair Interval—1993

Carrier	Avg. Missed Installations (Days)	Avg. Repair Interval (Hours)
Ameritech	5.0	2.3
Bell Atlantic	4.7	1.9
Bell South	3.7	4.4
NYNEX	4.2	5.9
Pacific Telesis	3.2	4.8
Southwestern Bell	4.0	2.8
US West	10.6	8.5
Contel	2.9	NA
GTE	3.0	6.2
United	NA	3.2

features a central network control center that monitors network performance 24 hours a day. End-users in buildings on the network can have new circuits activated in 24 hours. In case of an outage, service restoral in 4 hours is guaranteed. In spite of these extra features, prices are 15% to 20% less than for current private line offerings.

7. END-USER VIEWS OF COMPETITION AND NETWORK QUALITY

Although end-users do not speak with a single voice, a spokesman[17] for a group of large telecommunications users has expressed his view as follows: "Based on their experience over the last twenty years, large users believe that competition is far superior to regulation as a means of satisfying their needs. Users, therefore, strongly support the introduction of local exchange competition wherever feasible."

Particularly as competition has shifted to direct marketing to corporate end-users, LECs and ALT/CAPs have both sought to position themselves as value-added service providers. Although comprehensive quantitative data are not available, IXCs and special-access end-users believe they have seen increased network availability and reliability, enhanced service responsiveness, and lower circuit prices as a result. They believe, based on these results, that increased competition in the Local Exchange has increased network quality for special-access circuits, and it promises to do likewise for other network service offerings as competition spreads to include them.

8. OTHER MEASURES OF NETWORK QUALITY

Network availability is a very basic indicator of network quality. More detailed indicators of quality include measurements such as bit-error rate (BER), errored seconds, and so on. Such quality measurements are of greater importance as networks are used for data transmission. Most carriers and ALTs quote BERs of 10^{-9} for fiber circuits. Teleport quotes 99.9% error-free seconds. Customers with critical data needs can obtain performance standard quotations from vendors, but such standards are not routinely published nor are actual performance figures given.

9. FIBER-NETWORK EVOLUTION

Initial fiber networks were quite simple. They were, in fact, pseudo-networks, composed of collections of asynchronous point-to-point links. A voice-grade channel entering the system passed through several stages of multiplexing before being transported over the fiber link as a broadband optical signal. At a node the process was reversed through several levels of demultiplexing. The single, voice-grade signal could then be terminated or rerouted through a digital cross-connect before being remultiplexed for further transmission. This requirement for multiple interfaces and back-to-back multiplexers at every node created many potential points of failure. With the introduction of SONET (Synchronous Optical Network), this situation is changed. The synchronization in SONET allows the identification of the bits associated with a selected, single-voice channel without demultiplexing the data stream. Among other things, SONET provides the control to "drop and insert" a single-voice channel in a broadband optical signal. Therefore, with SONET, true fiber-optic networks can be created.

SONET technology found its first application in path-switched, self-healing ring structures. In this direct substitution for asynchronous links, no complexity is introduced. Traffic is still moved essentially between two points, and if the link is disrupted in the primary path, the entire signal is transmitted by the secondary path. SONET improves the ring performance only by reducing the number of multiplexing interfaces. Bellcore establishes a reliability standard for a 10-mile interoffice SONET trunk[19] of 4 min per year (99.9992% availability).

In more advanced applications, SONET rings are line switched rather than path switched. If a channel fails, the signal is looped back at the network node. This makes more efficient use of the network bandwidth and allows network structures that are not point-to-point rings but that may have distributed endpoints and may be a mesh rather than a ring. Network management and operations support systems standards for fully supporting such complex networks have not been developed. Carriers have implemented SONET in various simple configurations with the intention of later upgrades to more sophisticated architectures. However, because these initial implementations lack general interoperability, they may represent barriers to future global network development and interconnection.

Widespread deployment of complex SONET mesh networks will present operational as well as technical challenges for LEC and ALT/CAP carriers with interconnected networks. Even basic timing signals, which are critical to successful SONET operation, raise the issue of maintaining master clock references across networks. Successful comprehensive operation of future interconnected networks will require a degree of cooperative interaction among competitors.

10. SUMMARY

Competition in local exchange services is at a very early stage. The impact on network quality, to date, has been limited primarily to dedicated access circuits in major urban centers. In these locations, high-volume end-users and interexchange carriers have experienced increased network quality as both LECs and

ALT/CAPS have competed for their business and have provided self-healing fiber transport. The loss of less than 1% of revenue has stimulated both tremendous LEC investment in upgrading network quality and sharply lowering prices for services subject to competition.

This competition is rapidly accelerating, and major capital investment in the deployment of fiber networks is underway by all participants. These facilities continue to incorporate more advanced technology, including SONET electronics, integrated network management, and complex, interconnected, multiple network architectures. In a fully competitive "network of networks" telecommunications environment, many issues of network performance will depend on close cooperation and coordination among competing network operators. Successful network interconnection will not be limited to physical linkage but will include network signaling and database sharing, as well as administrative and engineering cooperation.

An early test of cooperative interaction among local telecommunications competitors came in the February 1993 bombing of the World Trade Center in New York. Response to this disaster in terms of maintaining service and rerouting circuits was greatly aided by the communication and coordination processes established just 1 year earlier by the New York Carriers Mutual Aid and Restoration Pact.

Attainment of a high level of operational cooperation among network operators in a fully competitive environment can be a more important factor than network technology and architecture in future network quality. Soft issues may be a larger determinant of network quality than hardware in the networks of the future.

ENDNOTES

1. Teleport Communications Group, *Teleport Report*, (1989), 7, 4.
2. Tomlinson, R. G. (1993). *1993 Local telecommunications competition . . . the ALT report*, II-8.
3. "Ex Parte Submission of Metropolitan Fiber Systems, Inc.," FCC Docket No. 91-141, May 27, 1992, Exhibit A.
4. Offering Prospectus, Intermedia Communications of Florida, Inc., April 30, 1992; and, Secondary Offering Prospectus, Intermedia Communications of Florida, Inc., Oct. 26, 1993.
5. Wilson, C. (1989). Hinsdale's Aftermath: COs at Risk, *Telephony*, (March 20, 1989), 21–25.
6. Tomlinson, R. G. (1992). *Alternate local transport . . . a total industry report*, 38.
7. FCC Docket 91-141 (1991).
8. Kraushaar, J. (1990). *Fiber deployment update . . . end of year 1989*, Industry Analysis Division, Common Carrier Bureau, FCC.
9. Kraushaar, J. (1991). *Fiber deployment update . . . end of year 1990*, Industry Analysis Division, Common Carrier Bureau, FCC.
10. Kraushaar, J. (1992). *Fiber deployment update . . . end of year 1991*, Industry Analysis Division, Common Carrier Bureau, FCC.
11. Kraushaar, J. (1993). *Fiber deployment update . . . end of year 1992*, Industry Analysis Division, Common Carrier Bureau, FCC.

12. *U.S. House of Representatives Review of Telephone Network Reliability and Service Quality Standards*, February 1992.
13. Bellcore (1992) "*Generic Reliability Assurance Requirements for Fiber Optic Transport Systems*," TR-NWT-000418, Issue 2.
14. See Note 13, Section 2-2.
15. MFS Press Release. (1992). *MFS urges the FCC and House of Representatives to raise U.S. network reliability standards*, April 8, 1992.
16. Kraushaar, J. (1993). *Quality of service for the local operating companies aggregated to the holding company level*. FCC Common Carrier Bureau, February 1993; and, Levine, H. D. (1993). A user perspective on competition in the provision of local exchange service, IRR Conference, San Diego, March 1993.
17. See Note 13, Section 2.2.4.

Chapter 10

Gigabits, Gateways, and Gatekeepers: Reliability, Technology, and Policy

John C. Wohlstetter
GTE Service Corporation

INTRODUCTION

Telephony met Information Age reality on January 16, 1990. It was AT&T's misfortune to lose over 50% of its network capacity when a single-bit "soft-glitch" cascaded through 114 SS7 adjunct processors in its 4ESS network—in 20 min. AT&T's SS6 traffic survived. (SS7 and SS6 are signaling protocols, i.e., rules governing network control information.) The impact software control of public-switched networks has had on network reliability became clear to all.

Even alone, the increasing dependence of networks on software control is cause for concern. Magnifying the danger, however, are the proliferation of diverse, computer-controlled customer premises equipment and, more significant, of increasingly interconnected, separately managed *networks*. Some of these issues were addressed by the FCC's Network Reliability Council in its June 1993 final report.[1] The Council's solid work—and the FCC's—have made a constructive contribution to improving network reliability, and both bodies deserve commendation. But some risks to reliability were neither fully resolved by the Council nor by the FCC itself. I propose to discuss one: policing software access to networks.

1. THE EMERGING META-NETWORK: FROM PHYSICAL TO VIRTUAL TELE-WORLDS

It is now common currency to call our public-switched network fabric a "network of networks," with linkage at both the hardware and software levels. Although it is true that even in the Age of Ma Bell there were hundreds of independent

225

company telephone networks, interconnected to the Bell System, in those days the collective whole was considered a unitary "national public-switched telephone network." It was, essentially, Bell-driven: based on Bell technology, under Bell standards, and pretty much playing by Bell rules. Today, physical network segmentation is a much broader phenomenon, bringing with it increasing software interdependence. We will find ourselves dealing with the consequences of this revolutionary paradigm shift into the next century.

A. Heterogeneous Hardware: LAN-CAP-IXC-Cell-LEC-LAN

Even the nation's first baby-boomer President can remember a time when phones came in three colors: basic black, midnight black, and pitch black. There was another side to this: One could have asked anybody at AT&T what kind of equipment was connected to the network and would have been told: black phones—by Western Electric. Of course, one could also have simply looked at the phone in the den.

This is, of course, no longer the case. Neither AT&T, nor the local exchange telephone companies, nor *anyone else* can say what is connected to the public network fabric today. What is connected behind the network demarcation point is, literally, *none of the network provider's business*. Customers now have desktop computers, mainframes, PBXs, FAX machines, hand sets—you name it—provided by hundreds of manufacturers scattered around the globe.

B. Seamless Software: My Bits . . . Your Bits . . . OUR Bits?

In a certain sense software represents a technological Faustian bargain: In exchange for a quantum leap in network capabilities—control, flexibility, new services—there is a troublesome price to be paid—the increased vulnerability of software-based networks. This vulnerability arises from four fundamental characteristics of network software: it is *global*; it is *programmable*; it is *accessible*; and it is *fragile*.

Global means that software represents a unitary logical overlay on dispersed physical network hardware. Thus, a single-point *logical* failure can, as happened to AT&T, cascade through dispersed physical nodes. *Hardware fails independently; no single-point hardware failure could have disabled half of AT&T's nationwide network capacity.*

Programmable means that software code can alter the way network hardware runs: Whereas picking up the telephone simply means closing an electric circuit between the phone and the central office, sitting at a PC the user can *redirect network assets*. Members of the hacker group "Legion of Doom" did just that a few years back, forwarding 911 calls in a Bell Operating Company's network to a dial-a-porn service.[2]

Accessible means that network assets are becoming more widely available, per Open Network Architecture (ONA). Service providers are gaining access to network software and pressing for complete control over the services they derive from telephone networks.

Fragile means that when software "breaks" it is not easy to "fix." It took AT&T 2 weeks to find the faulty code that brought its SS7 network down. They found an AND condition in place of an OR condition—out of *millions of lines of code.* Looking for this stuff is not made any easier in that at the start of the search one does not know what kind(s) of logical code error(s) one is looking for.

Now, add in multiple networks and multiple providers. The rash of SS7 network crashes in the summer of 1991 was caused by faulty code in an update of SS7 software provided by one vendor; companies not using that code were spared. It is only a matter of time before faulty code crosses a network gateway—to crash someone *else's* network.

C. Gateway to the Stars: A "Virtual Bridge" Entrance?

There will be more to say about this later, but for now simply note that at the entrance to each provider's network is a *gateway* that establishes, so to speak, the *rules of the road* for accessing the network. *Inherent in the nature of software is the ability—unless controls are effective—to reach across gateways and control the operation of distant networks.*

Technology is transforming today's networks: The central office switch is a digital computer; every desktop workstation or home PC is potentially a digital switch. Thus, transmission, switching, computer processing, and memory management functions, to date essentially distinct operations, are now being woven into a web of interconnected computing/communication networks.

The merger is a product of the combination of digital electronic hardware and software: Dispersed physical assets are controlled by a unitary overlay logical network. The logical overlay not only controls the operation of the physical network *infra*structure, it creates a functional *super*structure; access to network software logic enables both network providers and network users to define new network configurations—*virtual* networks. (In techno-parlance, *virtual* denotes the logical, software-defined equivalent of physical hardware functionality.)

II. REGULATING RELIABILITY: FROM HIPPOCRATES TO PANGLOSS?

When the nationwide network was primarily entrusted to AT&T—Theodore Vail's "one system, one policy, universal service"—Ma Bell guarded it as a national treasure. Any act that could conceivably bring harm to the network was simply *verboten.* Subscribers either took service on AT&T's terms or wrote letters. This began to change with equipment deregulation.

A. Harmless Hardware: The Legacy of Part 68

When the FCC began weighing rules to govern interconnection of equipment to telephone networks, AT&T, as part of its case in opposition, warned that if defective equipment was connected, harmful voltage—potentially lethal—could be sent over the network. Callers injured during a thunderstorm by lightning voltage could attest that the danger was not merely hypothetical.

Once it became clear that interconnection was inevitable, the debate shifted to what safeguards should be adopted and who would have responsibility. Equipment vendors denied that their equipment would cause harm and placed responsibility on the network provider. In the end, the FCC adopted Part 68, providing for interconnection on demand for equipment registered under Part 68. In doing so, the FCC in effect followed the precept of the legendary father of medicine, Hippocrates: "First, do no harm."

But Part 68 also enshrined another precept, for once and for all: Beyond the network demarcation point—in most cases, an RJ 11 modular jack—*what the customer does on the premises is—at least, generally—no one else's business.*[3]

B. Safe Software: The Promise—Hope?—of ONA

Open Network Architecture represents, essentially, the software equivalent of hardware interconnection. Just as the physical assets of the network were opened up, now the logical assets are opening to outside access. But there is a crucial difference: Hardware access means passive acceptance of network service; software access means potential *control* over network assets. The customer who merely connects equipment under Part 68 cannot redirect 911. Now this is changing—radically.

ONA is opening networks up to a potentially vast pool of users. With more people enjoying access to network features and with more of the network's innards (software primitives) being made available, opportunities for abuse—accidental or premeditated—of network assets will clearly increase, unless adequate countermeasures are implemented. The moral is: Unless we, like Voltaire's Pangloss in *Candide*, believe this to be "the best of all possible worlds," we need the equivalent of a software Part 68.

III. RELIABILITY AND RESPONSIBILITY: AM I MY TELE-BROTHER'S TELE-KEEPER?

As I briefly noted earlier, network entrances—*gateways*—represent the ports of call for information traveling through the network fabric. Increasingly, in a digital environment, all that the gateway will mark will be bits—an increasingly seamless, endless digital bit-stream: not voice, not data, not image, not video; just . . . BITS.

A. Gateways: Toll Booths on the Information Superhighway

Everyone who travels America's highways knows that, sooner or later, there will be tribute rendered to Caesar. The toll booth is as much an image of the automobile age as are tail fins. A network gateway can represent the same thing on Vice-President Gore's Information Superhighway: collection of necessary tribute to support the fabric. The toll paid is, of course, for exercising the right of access to network facilities. But is this enough?

B. Gatecrashers: Digital Dillingers, Accidental Tourists

Everyone has their own list of whom they consider yesterday's heroes. Some of mine include: Alexander Graham Bell, Theodore Vail, Edwin Armstrong, Claude Shannon, John von Neumann, and Robert Noyce.[4] But how many of these names ring a bell: Robert Tappan Morris, Pengo, Frank Darden? They are stellar attractions in an Information Age rogues' gallery: computer hackers cruising the information highway in search of prey.

Morris launched the INTERNET "worm" on its not-so-merry way one fine day in 1988, crashing 6,000 computers and causing, by one estimate, $98 million in lost computer time.[5] Pengo was a member of the West German hacker club, KAOS, which in 1987–1988 prowled through confidential Pentagon databases in search of information for the KGB.[6] And Darden was a member of the teen hacker group "Legion of Doom," whose rerouting of BellSouth's 911 service was a major telecaper. The first was a negligent prankster; the second, a spy; the third, a malicious prankster. They are part of the Information Age future. And we had better learn how to deal with them.

In addition to the "digital Dillinger" threat there is the problem of the "accidental tourist." The SS7 failures that crashed several local exchange carrier networks in the summer of 1991 were caused by a faulty software upgrade supplied by a single vendor of SS7 software. That vendor supplied SS7 software for 100 Signal Transfer Points (STPs) in several carrier networks; 57 STPs had the defective code installed.[7]

According to the FCC's own report on the STP failures, the outages were caused by a confluence of three factors: (a) three bits of faulty code supplied by the vendor, (b) a "triggering event," and (c) weekday "busy-hour" call overflow between 11 A.M. and 2 P.M.[8] The triggering events differed with each outage, but the common result was call congestion overflow on STP links. The vendor did not fully test the updated code.[9] Even had the code been thoroughly tested, the vendor conceded that it could not have simulated "a complete range of potentially contributing trigger sources."[10]

C. Gatekeepers: Toll Collectors or Bit Bouncers?

This is, so to speak, "where the rubber meets the road," where a software Part 68 would have to fit. Just as standards were adopted for registering hardware that is connected to the public network fabric, there now needs to be software standards.

Fixing responsibility for "bit bouncing" on gatekeepers is not an abstract issue. Last session of Congress saw legislation introduced that would have imposed financial penalties on carriers whose networks went down. The measure of damages would have depended on the scope and duration of the outage and the degree of fault assigned the carrier.[11]

Now, suppose that I tape my password to my PC, or that my password is "password." Someone logs on (either on premises or remotely) to my PC, and after entering the correct log-in name and password, is *for all intents and purposes a legitimate user*. Newly legitimized, the hacker now dials out through the office PBX and calls a network database in California. Bypassing security at the database—for example, by stealing passwords as Pengo did when entering some 400 military networks—the caller now sends to the database a little surprise: *Michelangelo*—the virus, not a video of the Sistine ceiling.

If the distant database is "zitzed out," who pays? *I* was negligent. Should Pacific Bell pay? As a common carrier with no right to control message content, PacBell merely carried bits over its network. *There need to be "rules of the road" that enable us to trace damage to the source and fix responsibility accordingly.*

Gateway policing is a software security issue that has been examined by the National Security Telecommunications Advisory Committee (NSTAC), a CEO-level body that advises the National Security Council.[12] In a 1992 report, NSTAC recommended that industry and government cooperate to develop uniform standards for public network—and internetwork—security.[13] The report concedes that "it is a leap to connect 'demonstrated collusion among hackers' to a group intent to take down the PSN."[14] But it concludes that a "serious potential threat exists: a resourceful adversary starting with the hacker information base."[15]

That base includes electronic bulletin boards—some with multilevel security so that top hackers can limit access to their purloined information.[16] More worrisome, the report notes a shift in hacker motivation toward "financial gain."[17] This contrasts with the traditional authority-defying motive. Finally, hackers have become more skillful at circumventing password protection and at defeating dial-back modem techniques.[18]

The report recommends possible action in six areas: (a) control of network element access (e.g., smart cards), (b) appropriate "level of suspicion" between networks (to isolate "weak links"), (c) recovery from software or database damage, (d) software memory partition and damage isolation, (e) network element analysis (e.g., audit trails), and (f) future architecture planning.[19]

D. Customers: A Tele-World "Reasonable User" Standard?

In March 1993, a Maryland federal court decided a suit brought by Jiffy-Lube International, a small business, against AT&T.[20] Jiffy-Lube sought reimbursement of $55,000 lost to a "call-sell" operator who successfully dialed into Jiffy-Lube's PBX. Calls were then made to the usual far-away watering-holes at Jiffy-Lube's expense. As articles in several national magazines have recently detailed, such "rogue resale" is on the rise.

Jiffy-Lube's claim ran head on into a contractual provision of AT&T's tariff, which held the "customer" liable for misuse. In granting summary judgment to AT&T, the Court gave short shrift to Jiffy-Lube's claim that AT&T should be held liable, despite the tariff provision, for carrying the hacker's call into Jiffy-Lube's PBX. Jiffy-Lube's case was not helped, one suspects, by their choice of password: "Lube." Nothing like originality.

So, given Jiffy-Lube's choice of password, would Jiffy qualify as a "reasonable customer"? Given widespread news reporting of hackers and call-sell rip-off artists, are not subscribers, with respect to their own network vulnerability, on "notice"—a legal term of art meaning what one *should* know, regardless of whether one *actually* does know? Should a "reasonable user" standard be more lenient for Aunt Tillie than for a Local Area Network manager? And if Aunt Tillie's teenager hacks from his PC, should she spot it?

E. Software Access: Who Gets to Play the Wizard?

Access to network software, the essence of ONA, can be understood at two distinct levels: *user-level* access, and *system-level* access. *User access* means the ability to avail oneself of network service applications; *system access* means the ability to *manage* network operations, that is, to change the way the network runs. A hacker's prime goal, upon entering a new system, is to become a *super-user* with all the powers of the system administrator (also called *administrative privilege*).[21]

System-level *access* thus means system-level *control*. As ONA service users— both competing network service providers and major users—penetrate deeper into the core networks of telephone companies, their access moves closer to the system-level line. They desire full software-based control over their network services, incorporating comprehensive functionality.

In pressing for deeper ONA, the November 1991 petition of the Coalition for Open Network Architecture Parties (CONAP) called for a "modular, transparent architecture."[22] Included in their concept of Open Systems Interconnection (OSI) is "access to system-level programs and commands."[23] They acknowledge the need for network security:

> No one would argue that the nation's public telephone network should be left "wide open" to anyone who might choose to wander into it; a high level of network security is an essential element of any public telephone network design.[24]

CONAP pointed to the "extreme success" of the open architecture adopted by IBM in the personal computer market. By analogy, they suggested that the telephone network should, increasingly, work just like a PC.[25] *Precisely*. Ask anyone whose hard disk has been "totaled" by some rogue program how safe computers are. In terms of economic impact, it is one thing to crash PCs and quite another to crash a central office switch.

It should also be noted that whereas IBM's open architecture has made IBM-compatible computers the most marketable, it has also made them the prime targets of the hacker community.[26] Apple's closed architecture has made its machines harder to penetrate. In noting this I do not intend to argue against open architecture per se, but merely to note a *collateral cost* of open access.

In such an environment, software "partitions" may be today's key line of defense. Hackers have, however, proven notoriously skillful at circumventing software-based defenses. Ultimately, hardware defenses may prove necessary.[27] *If in the meantime, network providers are required to open their system-level access, liability for harm should be shared.* One telecom consultant associated with the FTS-2000 contract stated:

> Networks are just an extension of the PCs, and virus protection should really begin at the terminal, regardless of the type of network you are using. If you don't stop the virus from getting into your PC, you won't keep it out of your network.[28]

For its part, the FCC has acknowledged that network reliability and integrity represent considerations associated with efforts of various interests to gain deeper software access to telephone networks.[29] A critical part of such an assessment is its apportionment of responsibility for harm done, just as is done with the equipment registration program.

IV. CYBER-CULTURE: WHO RULES "CYBERSPACE"?

Marshall McLuhan's "global village" is here—lest *anyone* doubt this a hacker in Melbourne, Australia was arrested in 1991 for breaking into American nuclear research and space agency computers, shutting down one Norfolk, VA NASA computer for 24 hours, altering and deleting data.[30] The village has a name—Internet—and already numbers millions of individual users. Streams of electrons and photons cross global network paths at warp speed. A New Yorker and a Malaysian communing via e-mail may share more in common than either does with their nextdoor neighbor. Electronic communities do not occupy land; they occupy what sci-fi writer William Gibson (in *Neuromancer*) named *cyberspace*. This did not signify much when telephone networks were radically different from their computer cousins. It does matter today. A new "cyber-culture" has emerged. For a moment, let's retrace its roots.

Historically, telephone and computer industry access/security cultures were diametrically opposite. For a century, telephone networks were closed systems, accessible by users almost exclusively for garden-variety voice communications usage. As recently as the mid-1950s, the old Bell System tried (ultimately un-successfully) to prevent customers from attaching a cup to the telephone, de-signed merely to allow users to converse privately in the presence of others (the "Hush-A-Phone" device). Deregulation, divestiture, and their twin offspring—equal interconnection and open access to network functionality—have radically altered the telephone industry culture.

Computing culture originally moved toward openness. In the early-1960s, computer use spread from a select few to university science campuses. Student programmers embraced a code of unbounded openness; computing creativity would be fueled by maximizing free access to systems and by programmers sharing their creative work with others in the computer community. The original cult of the computer hacker had as its hero the student prankster who would leave a humorous message on someone else's presumably inviolate machine. Hacking was also a way to help debug program code.

Three 1980s phenomena transformed the open computer culture. First, the explosion of the computer market, triggered by the success of the PC, made software vastly more commercially valuable than ever before and thus in need of protection from damage and piracy. Second, the rise of the malicious hacker, with an arsenal of "viruses," "worms," "time bombs," "logic bombs," and "Trojan Horse" programs,[31] made intrusion no longer the prankster's harmless high jinks. Access became a double-edged sword. Third, the rise of networking radically leveraged—for worse—the vulnerability of computers.

In a 1991 report, the National Research Council, operating arm of the Na-tional Academy of Sciences, appraised the risk of "soft terror":

> The modern thief can steal more with a computer than with a gun. Tomorrow's terrorist may be able to do more damage with a keyboard than with a bomb. . . . *To date, we have been remarkably lucky.* . . . As far as we can tell, there has been no systematic attempt to subvert any of our critical computing systems. *Unfortu-nately, there is reason to believe that our luck will soon run out.*[32]

Ironically, it was just as the telephone network was being opened up via Open Network Architecture that the computer world began to reexamine its own culture after the Internet debacle.

A. Cyber-Follies: 800, 900, 911, and 976

Mass announcement numbers pose hazards that network designers never antici-pated—indeed, even if they did it is doubtful if network economics would permit deployment of vast excess capacity that lies largely unused. (Historically, network design capacity has been determined by matching a desired blockage target—say, 1% of call attempts failing to gain access to the central office switch in the caller's exchange area—to traffic engineering statistical data that predict call blockage levels for a given number of lines serving a given number of customers.)

Ultimately, network economics may provide sufficient capacity to accommodate mass announcement calling without disrupting normal usage—when we enter our Fiber Future—but until then it is a live issue.

Already, mass announcement services have caused problems. In 1992, call-ins for tickets to hear music icon Garth Brooks jammed two local phone networks. One case was no laughing matter: A woman claimed that she could not reach 911 when her husband had a heart attack. Whether help would have arrived in time even with 911 cannot be said,[33] but the message is clear: 911 access must be safeguarded. The FCC has already acknowledged as much when it prevailed upon Pepsi to withdraw an 800-number call-in for the 1991 Super Bowl—on the eve of Desert Storm.

B. "Cyberpunks": Michelangelo Ex Maleficia

Every time I sit at my PC I thank the Lord that "Saddam don't know software." So far, at least that we know, terrorists seem to prefer buckets of blood to evil electrons.[34] Most hacking to date has been mere "cyber-pranks." *We cannot assume that we will continue to enjoy virtual immunity from software invaders who intend—and know how to inflict—real damage.*[35] Knowledgeable programmers who examined Robert Morris's code stated that had Morris wanted to destroy vast reams of Internet data, he need only have added a few lines of code to his worm—a task easily within the competence of Morris, a highly regarded UNIX programmer.[36]

The *Michelangelo* virus that destroys data on a PC hard disk can also destroy an SS7 database. ONA will require software "firewalls" to guard access. As outside access goes deeper into the core software network, the risk of compromise will surely increase. The battle here is no different than the classic match-up of armor and shell, which began when Hector's spear pierced but five out of the seven ox-hide folds of Ajax's shield. (The gods saved Hector to die another day. There may not be recourse to divine antiviral intervention.)

C. Cyber-Law: Cyber-Crimes and Tele-Torts

The Internet disaster prompted a rash of stricter laws to punish abuse of computer networks. Morris himself received a suspended sentence—his act was, after all, not the culmination of a career of malicious hacking, but rather a college kid's surrender to a spur-of-the-moment antisocial impulse, albeit causing huge financial harm.

As the network becomes more like a single, vast computer metanetwork, the problems that plague the computer world are bound to intrude into the telecom world. *Wilkommen, bienvenue, welcome*: viruses, Trojan horses, worms, bombs, and whatever else might be conjured up on the Island of Dr. Moreau. The dark side of the virtual tele-world is here.

To *cyber-crimes* must be added *tele-torts*. Those who use telephone networks to impair the reliability of the nationwide public network fabric must be held responsible. (Because most hackers do not have what the law calls "deep pockets"—and could not afford to purchase multimillion-dollar insurance policies to cover potentially vast network damage—criminal prosecution may be necessary in serious cases.) The FCC—and the states, for their part—should adopt "rules of the road" governing those who seek access to network software, to minimize the danger of network software being manipulated by hostile users. Remember: *Open access for the pharmacist is also open access for the drug dealer.*

What makes matters urgent, in this observer's view, is that global software transparency *raises* the potential payoff to software Darth Vaders—the damage from single-point failure is global. Nor can one count on user security alone: *Just as a secret is as safe as the biggest gossip that knows it, a network is as secure as its most careless user.*

A 1989 report by the National Research Council assessed the FCC's ONA policy and recommended: "At minimum, the evolution of ONA should reflect security considerations as well as the desire to provide open, equal access for users."[37]

Open networks are a necessity if the benefits of the Information Age are to be realized. But no more than any one would leave the front door open should network providers be required to do so. *Open networks must become open secure networks.*

The equivalent of a Software Part 68 is needed to address the range of technical and policy issues posed by potential abuse—accidental or intentional—of critical network software. At minimum, there need to be standards for testing, certification, and registration of software, calibrated to authorization levels—with secure "firewalls" separating user- and system-level access.[38] It will be necessary to coordinate any FCC action with ongoing activities of the NSTAC.

The NSTAC should continue its fine work in assessing software security threats and coordinating industry/government responses. The FCC should explore issues pertaining to legal responsibility and public policy. It should examine the relative responsibility of vendors, service providers, and common carriers, reconciling open access with network integrity and security. It should consider: (a) what knowledge, if any, a "reasonable user" should be deemed to have legal notice of; (b) possible testing, certification, and registration regulations; and (c) working with industry to develop standardized tools, such as audit trails, to help fix responsibility for network harm.

Responsibility must follow control. Where control lies, so lies responsibility. Those who link software to the core network should accept the same obligation imposed upon those connecting hardware: "First, do no harm." *Ease of access and ease of security are flip sides of the same coin; access without restriction is access without security.*

Everyone is familiar with three "famous last words": "The check is in the mail"; "Of course I'll respect you in the morning"; and "Hi! I'm from the IRS and I'm here to help you." In a software-driven world, a fourth can be added to the list of classics: "Relax! This software is completely bug-free and absolutely secure." We discount software risks at our peril.

ACKNOWLEDGMENTS

This chapter is based on a talk that was originally presented April 23, 1993, at a seminar, *Quality and Reliability of Telecommunications Infrastructure*, at the Columbia Institute for Tele-Information. Its contents are entirely attributable to the author and in no way constitute any position(s) taken by the GTE Corporation, or by any of its affiliates or subsidiaries, or by any officer or employee thereof. The author assumes full responsibility for the contents, including any errors.

ENDNOTES

1. *Network Reliability: A Report to the Nation*, Federal Communications Commission's Network Reliability Council (National Engineering Consortium, 1993).
2. On July 9, 1990, three members of the Legion pled guilty to federal fraud charges in Georgia. *Telecommunications Reports*, July 16, 1990, p. 27. But the Legion's 911 caper ended in an anticlimax, as it turned out that the information necessary to access the 911 software, which a Legion member had broadcast over an electronic bulletin board, was also available from Bellcore via an 800-number, leading to some charges being dropped against members of the group. *Communications Daily*, July 31, 1990, pp. 2–3. The event did, however, show that 911 software was manipulated from outside. "Open Sesame: For Hackers Such as Frank Darden, There's Nothing More Inviting Than a Closed Door," *Wall Street Journal*, August 22, 1990, p. 1.
3. This "none of the phone company's business" precept represented a 180-degree shift from olden days. The Bell System's historic position had been that no one could attach *any* device to Bell equipment without the company's express permission. Prodded at times by the courts, the FCC demolished Bell's position and devised rules governing attachment of non-Bell equipment to the network, culminating in adoption of the Part 68 regulations in the mid-1970s. An excellent account of the regulatory and legal twists and turns over a 30-year period can be found in Kellog, Thorne, and Huber, *Federal Telecommunications Law*, pp. 499–509 (Little, Brown & Company, 1992).
4. Vail: the architect of the modern Bell System; Armstrong: America's radio genius, invented the superheterodyne receiver and FM transmission and recognized by his peers (but not the courts) as inventor of the triode vacuum tube; Shannon: the father of information theory; von Neumann: the father of the modern electronic digital computer; Noyce: co-inventor of the integrated circuit and founder of Intel Corporation. Bell? Your guess.
5. The estimate comes from the Computer Virus Industry Association (San Jose, CA). McAfee, John, *Computer Viruses, Worms, Data Diddlers, Killer Programs, and Other Threats to Your System*, pp. 4 & 7 (St. Martin's Press, 1989).
6. Hafner, Katie and Markoff, John, *Cyberpunk: Outlaws and Hackers on the Computer Frontier*, pp. 139–251 (Simon & Schuster, 1991).

7. *Preliminary Report on Network Outages*, p. 8 (Common Carrier Bureau, July 1991). The report, albeit labeled "preliminary," was the only FCC document issued on these STP crashes. It thus stands as the FCC's "final" statement on the matter.

8. *Ibid.*, p. 5. The SS7 software vendor was DSC Communications Corporation.

9. *Ibid.*, p. 1.

10. *Ibid.*, p. 8.

11. HR 4343, introduced by Rep. Edward J. Markey (D-7, MA) in the 102nd Congress. The bill did not reach the markup stage. It has not been reintroduced, and no bill setting penalties is pending at this date.

12. The NSTAC was formed in 1982 to address national security emergency preparedness (NS/EP) issues in light of the AT&T divestiture. NSTAC works closely with the National Communications System (NCS), established by President Kennedy in 1963 (after the Cuban Missile Crisis revealed a need for better crisis communications). NCS is part of the Defense Information Systems Agency. The NSTAC has considered software security issues, and continues to do so.

13. Final Report of the Network Security Task Force (NSTAC, June 10, 1992). The task force is a subgroup established under the aegis of the NSTAC's Industry Executive Subcommittee (IES).

14. *Ibid.*, p. 17. (Emphasis in original.)

15. *Ibid.*

16. *Ibid.*, p. 18.

17. *Ibid.*, p. 17.

18. *Ibid.*, pp. 18–19.

19. *Ibid.*, pp. 11–13.

20. *AT&T v. Jiffy-Lube International, Inc.*, CIVIL NO. R-90-2400 (D.D.C., Md., filed February 18, 1993).

21. Landreth, Bill, *Out of the Inner Circle*, pp. 51–52 (Microsoft Press, 1989). A former hacker, Landreth disclosed techniques he used to gain "super-user" status. He was ultimately caught by the FBI and tried. After conviction, he was sentenced to community service and a small fine.

22. Petition for Investigation, Coalition of Open Network Architecture Parties (filed November 16, 1990). CONAP's petition led to issuance of the FCC's Notice of Inquiry in CC Docket No. 91-346 (note 29 *infra*).

23. *Ibid.*, p. 30.

24. *Ibid.*, p. 32.

25. *Ibid.*, pp. 7–8.

26. According to one 1989 estimate, 70% of viruses struck IBM or IBM-compatibles, compared to 24% for Macintosh and Amiga systems, and 6% for all others. McAfee, note 5 *supra*, p. 60.

27. See generally, Hoffman, Lance J. (Ed.), *Rogue Programs: Viruses, Worms and Trojan Horses* (Van Nostrand Reinhold, 1990).

28. *Ibid.*, p. 300, quoting a consultant for Centel's FTS-2000 bid.

29. *In the Matter of Intelligent Networks*, Notice of Inquiry, CC Docket No. 91-346 (released December 6, 1991).

30. "Australia to Try Computer Hacker Accused of Damaging NASA Network," *Washington Post*, August 15, 1991.

31. These terms are frequently conflated into the generic designation *virus*—adequate for public discourse if irritating to the purist. The proper definitions are: *Virus*: Program code embedded within a host program that can only be activated by execution of the host and replicates itself into other hosts (e.g., the Pakistani Brain, which infects floppy diskettes). *Worm*: An independent program that can execute and replicate itself, without prior execution of another program (e.g., the Morris Internet worm, which clogged computer memory). *Trojan Horse*: A malicious program concealed within a legitimate program (e.g., the "Sexy Ladies" program that erased sectors on Apple Macintosh hard disks while their users admired the screen display). *Time Bomb*: A program triggered by occurrence of a temporal event (e.g., Michelangelo, activated on March 6, the artist's birthday). *Logic Bomb*: A program triggered by occurrence of a logical

condition (e.g., certain keystrokes or commands; in 1988, an employee of a Fort Worth insurer/brokerage firm left a rogue program on the system of his former employer, USPA & IRA Company, which was triggered when his name was removed from the payroll list). Hoffman, note 27 supra, pp. 23–25 and 205.

32. National Research Council, Computers at Risk: Safe Computing in the Information Age (National Academy Press, 1991), p. 7.

33. "Phone Tie-Up Blocks Aid for Dying Woman," Washington Times, July 24, 1992, p. C5.

34. Some terrorists apparently do. The PLO Virus was implanted in computers at Israel's Hebrew University, to be activated on May 13, 1988, the 40th anniversary of the last full day of juridical existence for the League of Nations' 1920 Palestine Mandate (the day before official proclamation of the formation of the State of Israel). The virus replicated too quickly and was discovered by Israeli technicians and neutralized before harm was done. See Fites, Philip, Johnston, Peter and Kratz, Martin, The Computer Virus Crisis, p. 30 (Van Nostrand Reinhold, 1989).

35. America's Hidden Vulnerabilities: Crisis Management in a Society of Networks, A Report of the Joint Panel on Crisis Management of the CSIS Science and Technology Committee (1984). The panel's co-chairmen were R. James Woolsey, now CIA Director, and Robert H. Kupperman, a noted terrorism expert.

36. Hoffman, note 27 supra, p. 221.

37. National Research Council, Growing Vulnerability of the Public Switched Networks: Implications for National Security Emergency Preparedness, p. 37 (National Academy Press, 1989).

38. In August 1993, the FCC issued a Notice of Proposed Rulemaking, in which it proposed "third-party mediated access" to network software. In doing so, the Commission gave explicit recognition to network reliability as a major public policy goal and conditioned system-level access to network software on a demonstration of genuine third-party need for access, ensuring that such access does not operate to jeopardize network reliability. Access would be implemented in three phases, with network reliability concerns having greater weight if access goes deeper into the network core. In doing this, the FCC's access criteria implicitly incorporate Part 68's "first, do no harm" principle. See CC Docket 91-346 (FCC 93-380, released August 31, 1993), note 29 supra.

About the Authors

Sanford V. Berg is a Florida Public Utilities Professor in the Department of Economics at the University of Florida and the Executive Director of the Public Utility Research Center. He teaches courses in introductory microeconomics, public policies toward business, and undergraduate and graduate courses on public utility economics. His fields of specialization and research include public utility economics and pricing policies as applied to telecommunications and energy, as well as joint venture activity and innovation, the impacts of technological change on regulation, and market organization. He holds a PhD and an MPh in Economics from Yale University and a BA from the University of Washington.

Bhaskar Chakravorti is a member of the technical staff in the Economics Research Group at Bellcore. His fields of interest are: game theory and the economics of information, industrial organization, public finance, economics of telecommunications, and information systems. He has published papers in a variety of academic journals, including the *Journal of Mathematical Economics*, *International Economic Review*, the *Journal of Public Economics*, *International Journal of Game Theory*, the *IEEE Transactions on Automatic Control*, and several others. He has been a consultant to several telecommunications companies, such as NYNEX, Southwestern Bell, and STENTOR (Canada). Bhaskar Chakravorti has a PhD in economics from the University of Rochester, an MA in economics from the Delhi School of Economics (India), and a BA in economics from St. Stephen's College, Delhi University (India).

Nicholas Economides is Professor of Economics at the Stern School of Business since 1990. His fields of specialization and research include industrial organization, oligopoly, economics of networks, especially of telecommunications and of information, economics of technical compatibility and standardization, and the structure and organization of financial markets. He has published widely in the areas of networks, telecommunications, oligopoly, differentiated products, and liquidity and the organization of financial markets and exchanges. He holds a PhD and an MA in Economics from the University of California at Berkeley as well as a BSc (First Class Honors) in Mathematical Economics from the London School of Economics. He has taught at Columbia University (1981–1988) and at Stanford University (1988–1990).

Thomas Hazlett is Associate Professor of Agricultural Economics and Director of the Program in Telecommunications Policy at the Institute of Governmental Affairs at the University of California, Davis. In 1991–1992 he served as Chief Economist of the Federal Communications Commission in Washington, DC, and in 1990–1991 he was a Visiting Scholar at the Center for Telecommunications and Information Studies at the Columbia University Graduate School of Business. His academic work focuses on government regulation and the economic analysis of legal institutions, particularly in the telecommunications sector. His books on radio spectrum allocation and on cable television regulation (coauthored with Greg Sidak and Matt Spitzer) are forthcoming from MIT Press in 1994. Hazlett received his doctorate in economics from the University of California–Los Angeles in 1984.

Raymond W. Lawton is Associate Director for Telecommunications Research at the National Regulatory Research Institute and Adjunct Associate Professor at the School of Public Policy and Management at The Ohio State University. He received his PhD from the Ohio State University. Dr. Lawton has specialized in telecommunications competition, regulatory reform, modernization, management audits, water policy, and qualitative research methods. He has regularly made presentations at national regulatory forums and conferences and is a member of the National Association of Regulatory Utility Commissioners Staff Communications Committee. He recently was a faculty member at a telecommunications tariffs seminar in Tashkent, Uzbeckistan, put on for the Confederation of Independent States sponsored by the U.S. Department of State.

William Lehr is an Assistant Professor of Finance and Economics at the Columbia University Graduate School of Business and a Faculty Research Associate at the Columbia Institute for Tele-Information. His fields of specialization and research include industrial organization, political economy, and regulation, especially as these apply to information technology industries. He teaches courses in microeconomics and competitive strategy, including a course on the media. He holds a PhD in economics from Stanford, an MBA from the Wharton Graduate School, and MSE, BS, and BA degrees from the University of Pennsylvania.

Jonathan M. Kraushaar is a staff engineer in the Industrial Analysis Division of the Federal Communication Commission (FCC) Common Carrier Bureau.

He has been engaged in a variety of regulatory activities involving AT&T and the Bell Operating Companies and has compiled FCC studies on fiber deployment activity, quality of service, and bypass. He has co-authored a number of technical papers, including several on telephone traffic engineering, and he holds three patents, two of which deal with telephone traffic measurement. Kraushaar has been instrumental in setting up the FCC-State Link electronic bulletin board at the FCC and now serves as its system operator. His current interests include activities associated with the Commission's ARMIS database, which houses data on the local telephone operating companies. Mr. Kraushaar received his MS and BS degrees in Electrical Engineering from Carnegie-Mellon University.

William Sharkey is currently a member of the economics research group at Bellcore in Morristown, New Jersey. Previously he has been a member of the technical staff at Bell Laboratories and held visiting appointments at several universities. His fields of specialization and research include game theory and its applications to microeconomics, cost allocation, the economics of networks, and the economics of telecommunications. He is the author of *The Theory of Natural Monopoly*, a chapter in the *Handbook of Operations Research* on "Network Models in Economics," and he has recently completed a series of articles on axiomatic methods of pricing and cost allocation. He holds a PhD in economics from the University of Chicago and a BS from the University of Michigan.

Yossef Spiegel is currently a member of the economics research group at Bellcore in Morristown, New Jersey. His fields of specialization and research include regulatory economics, industrial organization, and corporate finance. He has recently completed a series of articles on the financial decisions of firms under rate regulation. He holds a PhD in economics from Northwestern University and MA and BA degrees from Tel Aviv University.

Padmanabhan Srinagesh is a Member of Technical Staff with the Information Networking Research Laboratory in the Applied Research Area at Bellcore. His current responsibilities include the economic analysis of government initiatives to develop an Information Superhighway and the analysis of evolving business relationships in the global Internet. He also works on a variety of pricing issues raised by the introduction of a common platform that can provide voice, video, and data services. Before joining Bellcore, Dr. Srinagesh taught economics and business courses at Williams College and the University of Illinois at Chicago. He has published articles in several journals, including the *American Economic Review*, the *Quarterly Journal of Economics*, and the *Journal of Regulatory Economics*. Dr. Srinagesh received his MA in Economics from the Delhi School of Economics in 1974 and his PhD in Economics from the University of Rochester in 1980.

Neal Stolleman is a Market Manager within the Systems and Operations Line of Business at Bellcore. He is currently focusing on how the demand for operating system software is influenced by the evolution of telecommunications networks toward broadband and PCS technologies. He has also developed PCS and residential video strategic planning models. Prior to Bellcore, Dr. Stolleman was Regulatory Policy Economist at GTE Service Corporation, where his research included alternative regulatory systems, telco/cable TV cross-ownership, bypass of the public switched network, and multioutput nonlinear pricing. Dr. Stolleman published "Economies of scope in the provision of narrowband and switched broadband services" in the January/February 1993 issue of *Telecommunications Policy* and has presented his work to the International Telecommunications Society and International Symposium on Forecasting, in the United States and Europe. He holds a PhD in Economics from the City University Graduate Center, New York.

Richard Tomlinson is the President of Connecticut Research, a consulting and publishing company that he founded in 1986. He publishes a monthly newsletter, *Connecticut Research Report on Competitive Telecommunications*, and is a frequently invited speaker on the status of competition in telecommunications. He has published numerous marketing studies, as well as over 35 technical papers, and holds five patents. He is a Senior Member of the Institute of Electrical and Electronics Engineers and holds three degrees in electrical engineering: a PhD from The Ohio State University and BS and MS degrees from Case Western Reserve University. He also holds an MBA from Rensselaer Polytechnic Institute.

John Wohlstetter is Director of Technology Affairs for GTE Corporation in Washington, DC. His primary responsibility is to assess the impact of government policies on the company's long-term network deployment strategy. Among the issues he covers are: broadband networks, HDTV, local loop competition, regulatory reform, and network reliability. Before joining GTE he was with Contel Corporation, beginning there as an attorney. From 1986 to 1989, he served as Senior Adviser to a panel convened by the National Academy of Sciences that examined public network vulnerability from the standpoint of national security preparedness. Mr. Wohlstetter holds a BBA in Finance from the University of Miami, a JD from Fordham University, and an MA Public Policy/Telecommunications from George Washington University.

Author Index

The letter *t* following a page number indicates tabular material.

Subject Index

The letter | f | following a page number indicates a figure; | n | denotes a footnote; and | t | indicates tabular material.